MYSTERIES

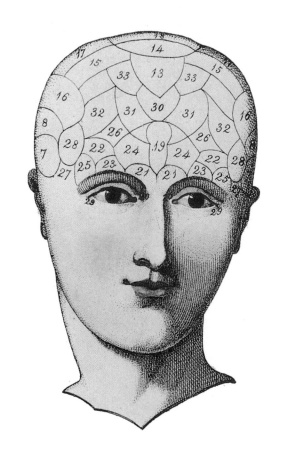

OF THE MIND

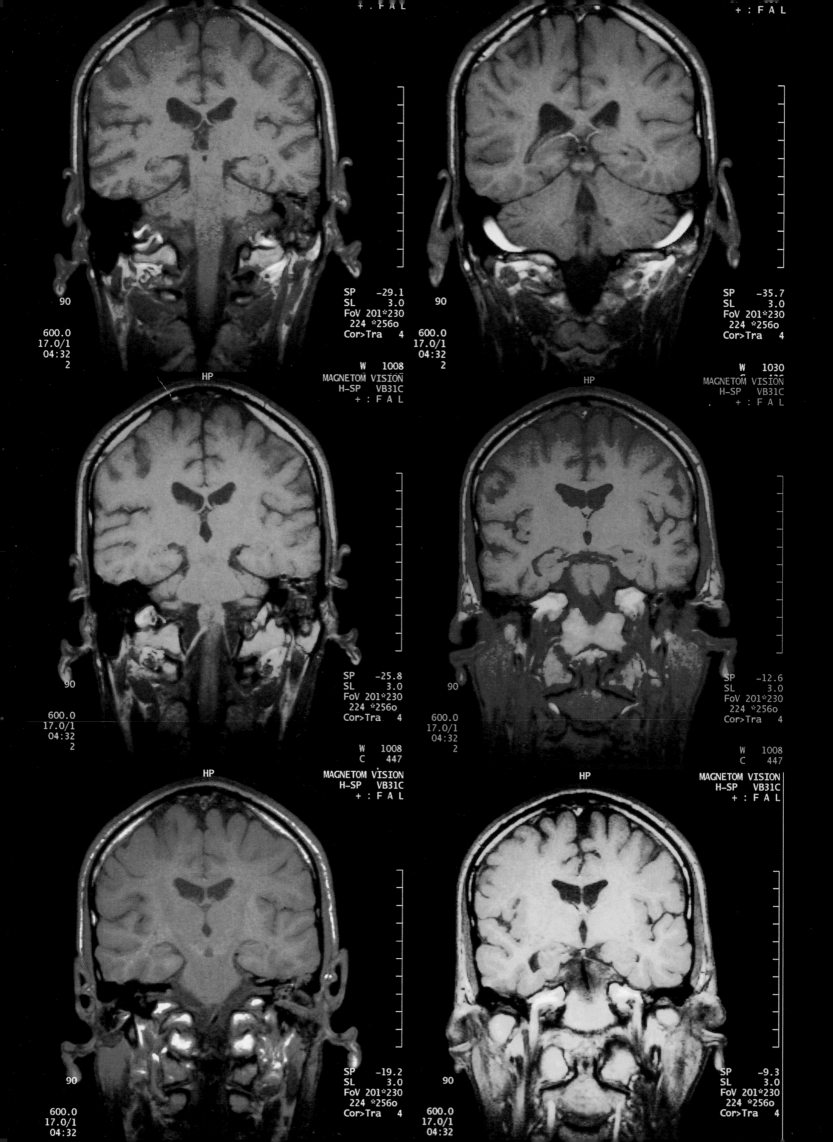

MYSTERIES
OF THE MIND

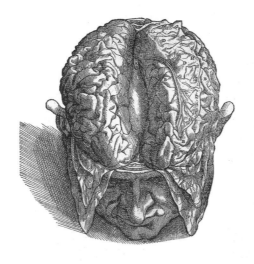

RICHARD RESTAK

**NATIONAL
GEOGRAPHIC**

Washington, D.C.

PRECEDING PAGE: Phrenological map devised partly by Franz Gall divides the head into numbered areas of function. OPPOSITE: Colorized MRI scans depict various "slices" of living human brains. ABOVE: A woodcut by Andreas Vesalius reveals the two cerebral hemispheres in all their glory.

CONTENTS

■ Small beginnings: Like this newborn, the human brain is far more complex than it first appears. Though it weighs a mere three pounds, it contains about 100 *billion* neurons, or nerve cells, each connected—directly or indirectly—to as many as 100,000 others. Establishing and maintaining such connections is the basis of human learning and development.

OF MIND AND BRAIN

Whenever we speak of the mind—making up our mind, improving our mind, changing our mind—we're actually referring to activities carried out by our brain. Thoughts, memories, feelings—all the various expressions of mind—necessarily involve brain activity. Indeed, "mind" is simply a commonly employed descriptive term for the operations of the brain.

Our present understanding of the relationship of mind and brain developed slowly, over many centuries. What's more, the theories that became popular at different times and places often mixed reason, fantasy, and speculation.

Plato relied upon geometry for his theory of mind. Since the sphere is the perfect geometric figure, he reasoned that the head and its enclosed brain provided the biological underpinning for mind. In contrast, Plato's preeminent pupil, Aristotle, attributed all life forces to the heart and considered the brain to be nothing more than a cooling system for the blood.

Many centuries later, William Shakespeare's *The Merchant of Venice* would echo the heart-brain dilemma concerning the basis for mind:

> *Tell me where is fancie bred*
> *Or in the heart or in the head.*

The Greek physician Hippocrates came up with yet another approach, suggesting that personality—one among many expressions of mind—depended on the preponderance within the body of the four humors (what we now identify as chemicals or bodily fluids). These were phlegm, yellow bile (also known as choler), the black bile of melancholy, and blood.

Even today, humoral terminology endures in the form of characterizations. We consider grumpy people to be in a bad humor. We also refer to the unemotional as phlegmatic, the confident as sanguine, the ill-tempered as bilious, and the emotionally explosive as choleric.

And although no one today seriously attributes moods to the existence of humors circulating in our bodies, scientists have found that various messenger chemicals known as neurotransmitters act within the brain to influence mind and personality. For instance, depression, schizophrenia, and bipolar disorder (also called manic depression) are all thought to result from imbalances of the brain's neurotransmitters.

This transition over the centuries from humors to neurotransmitters exemplifies a basic principle: rather than disappear entirely, theories about the mind and brain tend to undergo reformulation and transformation.

For instance, scientists once debated which was more relevant to the operation of the brain, the actual brain tissue that can be seen and palpated—or the fluid-containing cavities, called ventricles, which lie deep within. While most of us today readily choose solid brain tissue over fluid-filled cavities, this conclusion took many centuries to evolve.

Greek physician Galen harbored no doubt that the substance of the brain was the seat of the "soul"—what we would now call the mind. To prove his point he carried out an experiment, cutting the motor fibers that extend from a pig's brain to its limbs, and the sensory nerves that convey sensations from limbs to brain. He concluded that the sensory

■ Mental giants: Plato and Aristotle discuss the nature of the mind in this stylized painting, *The School of Athens*, by Raphael.

nerves traveled to the front of the brain, which he considered soft and thereby better able to receive impressions, while motor impulses originated further back in a firmer area of the brain. (Actually, Galen had things reversed; the motor centers are towards the front of the brain, while the sensory centers are toward the back.)

While he muffed a few important details, Galen was right on target about the centrality of the brain. In the words of his role model, Hippocrates, "Not only our pleasure, our joy, and our laughter but also our sorrow, pain, grief, and tears arise from the brain and the brain alone. With it we think and understand, see and hear, and we discriminate between the ugly and the beautiful, between what is pleasant and what is unpleasant, and between good and evil."

About 1,200 years after Galen, Leonardo da Vinci carried out an ingenious series of dissections that definitively established the importance of brain tissue. He poured molten wax into the ventricles of the brain of an ox, let it set, and then dissected and examined the surrounding brain tissue. He discovered that sensory nerve fibers terminate in the substance of the brain itself, in an area later named the thalamus.

A contemporary of Leonardo, the anatomist Andreas Vesalius, provided further proof of the centrality of the brain over the ventricles by studying the brains of freshly executed criminals. Typically the heads, still warm and dripping blood, were brought to Vesalius for dissection and illustration. With the assistance of artists from Titian's studio, he created 15 copperplate etchings that have been called "among the most outstanding drawings in neuroanatomy ever produced." On the basis of his dissections, Vesalius dismissed the importance of the ventricles, considering them merely "cavities and spaces in which the inhaled air…is, by the power of the peculiar substance of the brain, transformed into animal spirit."

But 17th-century philosopher René Descartes took a different slant. Observing that the brain collapses and rapidly deteriorates after dissection, Descartes proposed that in the living brain the ventricles served as a receptacle for a fluid that he felt was the unit of nervous transmission. He called this fluid the "animal spirits," and believed that the brain distributed it through the nerves that extend throughout the body. Within those nerves, he reasoned, tiny "valvules" controlled the inflow of animal spirits. For example, moving one's hand near a hot fireplace would stimulate skin receptors to pull on special filaments that opened a valvule in the ventricles. This action released the animal spirits into the nerves which, in turn, activated the muscles to move the hand away from the fire.

This so-called reflex theory explained involuntary actions such as automatic withdrawal from a flame. In contrast, voluntary or "willed" behavior—such as deciding to light the fire in the first place—demanded the additional participation of what he called the "rational soul"—what we would now call the mind. To Descartes the body was a machine, an automaton that took orders from the soul or mind. But how were these orders delivered? And where were they processed and ultimately carried out?

Descartes suggested that the interaction of the mind and brain took place within the pineal gland, a small central structure suspended close to the

Considered the "Father of Medicine," Greek physician Hippocrates (far left) recognized the importance of the brain as the organ of mind. Scientist and artist Leonardo da Vinci (left) used wax to make the first cast of the brain's inner ventricles.

ventricles and thus surrounded by the reservoir of animal spirits. Today, some 400 years later, the pineal has been demoted from the "seat of the soul" to an organ that secretes melatonin, a chemical associated with darkness and the induction of sleep.

Even so, Descartes remains famous today—not for his theory about the pineal, but for his insistence on the separation of brain from mind, a theory referred to by philosophers as mind-body dualism. Opposed to this is the belief that the mind has no physical existence but is best thought of as a functional term that we employ when referring to operations of the brain. This issue remains far from settled; controversy continues today in some quarters about the mind-body problem, more accurately described as the mind-*brain* problem.

A contemporary of Descartes, Thomas Willis, divided the brain into separate regions. Mental functions such as memory, will, and imagination could be localized, he believed, to specific brain areas. The renowned draftsman and architect Christopher Wren, who later designed St. Paul's Cathedral in London, provided illustrations of these areas in Willis's book, *Cerebri Anatome* (Anatomy of the Brain). Although neuroscientists no longer believe in the strict subdivisions espoused by Willis, his claim that discrete functions are localized within the brain led to the development of the medical specialty of neurology. In fact, Willis contributed the word "neurology" to medical science, and the arteries at the base of the brain are still known as the circle of Willis, due to his original description.

But the path from Willis to a modern understanding of the brain involved several detours and blind alleys. German physician and anatomist Franz Gall (1758-1828) established the pseudoscience of phrenology. Taken from the Greek words for "mind" and "discourse," phrenology involves evaluating a person's character from bumps and other irregularities of his skull. Part scientist, part charlatan, and all showman, Gall claimed exaggerated and often ludicrous correlations between the shape of the skull and personality characteristics. He maintained, for instance, that a person's tendency for secretiveness or conscientiousness or sublimity could be assessed simply by palpating the skull and then correlating the findings to one of the many ever-changing phrenological maps.

Despite such superficial and often bizarre claims made by Gall and his many followers, phrenology did set the stage for more scientific correlations of brain, mind, and behavior. For one thing, it provided an alternative to the mind-brain dualism espoused by Descartes. More importantly, it served as a forerunner of the modern belief that mind is an expression of brain structure and function. In addition, phrenologists suggested the first theory of localization—the concept that the brain can be divided into areas that carry out specialized functions. They also were the first to establish that the brain's gray matter was composed of brain cells—neurons—while white matter consisted of fibers connecting neurons to each other.

Pierre Jean Marie Flourens, a leading French physiologist in the 19th century, vehemently opposed phrenology and depicted Gall as a madman obsessed with the collection of human skulls. As an alternative to localization, Flourens stressed that the brain must be considered a single working unit rather than a collection of separate independent

Philosopher René Descartes (far left) held that the conscious mind was distinct from the bodily machine. A diagram of his reflex theory (left) depicts how "animal spirits," which he believed were conveyed through the body by a hydraulic system, prompted a hand to withdraw from a flame.

centers. He arrived at this belief, known as holism, after performing a series of rather grisly experiments on birds, mice, cats, and dogs. Flourens noted that he could cut out large portions of these animals' brains without causing any observable change in their behavior. This led him to declare that vision, hearing, and other sensory and motor processes were not localized in separate brain regions but were distributed over large cortical areas.

Who was correct, Gall or Flourens? Is the brain organized with specific areas separated from each other like countries depicted on a map? Or is the situation more like a large and intricately linked lattice network? In truth, both ideas about the brain are supported by evidence. For instance, a stroke affecting the left hemisphere often leads to speech and language impairment, which suggests the left hemisphere is specialized for processing language. Yet the brain also functions holistically. Sights, sounds, and other sensory experiences don't exist in isolation but are synthesized within the brain into a single integrated experience.

Eventually a compromise was reached; most 19th-century scientists concluded that the simpler brain activities, such as movement and the elementary sensations of vision and hearing, are localized in specific brain areas, while most of the so-called higher mental functions—reasoning and memory, for example—are more diffusely distributed. Yet sometimes drawing a line between localization and holism stymied even the experts. The most famous instance occurred on September 13, 1848—a day that marked a watershed in our understanding of the human brain.

On that date an explosion at a railroad work site in Cavendish, Vermont, sent a thick steel rod through the head of foreman Phineas Gage, severely injuring the frontal lobe of his brain. Miraculously, Gage survived; his head slowly healed and he retained all his mental faculties and senses. But soon friends noticed that, although his abilities to add figures and speak and reason seemed unchanged, his personality differed dramatically from what it had been. Formerly temperate, restrained, and reliable, Gage became (in his doctor's words) "fitful, irreverent, indulging at times in the grossest profanity, manifesting but little deference for his fellows, impatient of restraint or advice when it conflicts with his desires, at times pertinaciously obstinate, yet capricious and vacillating, devising many plans for future operation, which are no sooner arranged than they are abandoned."

The French physiologist Flourens, eager to defend his holistic theory, ignored Gage's altered personality and declared the railway foreman unaffected by his injury. Gall, who died years before Gage's accident, undoubtedly would have considered the behavioral changes as proof that higher mental functions were localized. What's certain is that the sad case of Phineas Gage stimulated neurologists and other neuroscientists to find evidence that supported either the holistic or the localizationist point of view.

Twelve years after Gage's injury, the anatomist and anthropologist Paul Broca encountered a stroke patient nicknamed Tan, because "tan" was the only word the patient could say after his stroke. When Tan died, Broca performed an autopsy and discovered evidence of damage to the left frontal lobe. That experience and subsequent studies of patients who suffered from language difficulties convinced Broca of the importance of the left hemisphere to

Nineteenth-century poster touts the pseudoscience of phrenology, popularized by Franz Joseph Gall, who taught that a person's mental characteristics could be determined by palpating and "reading" the physical irregularities of his skull.

language. But if language processing is distinct to the left hemisphere, what functions does the right hemisphere carry out? That question would take a hundred years to answer.

In the 1960s, neuropsychologist Roger Sperry of the California Institute of Technology demonstrated that each cerebral hemisphere is specialized along certain lines. His research involved epileptic patients who had undergone a drastic surgical operation—the severing of the corpus callosum, a bridge of neural fibers connecting the two cerebral hemispheres—in order to control their epilepsy. The patients no longer suffered seizures, since the operation prevented seizure discharges from crossing from one hemisphere to the other. But because the corpus callosum had been cut, each cerebral hemisphere of these "split-brain" patients operated in comparative isolation from the other.

Sperry carried out a series of innovative, Nobel Prize-winning experiments aimed at finding out what each hemisphere does best. He discovered that the left hemisphere excels in reading, writing, and processing information in a sequential fashion. The right hemisphere is specialized for recognizing faces, drawing or reading maps, solving jigsaw puzzles, and expressing and perceiving the emotional nuances accompanying verbal and facial expressions.

While these findings in split-brain patients are intriguing, they must be put in perspective: In those of us with an intact corpus callosum, the two functionally asymmetric hemispheres work together in unity. Thus Sperry's split-brain work is consistent with *both* the localizationist and the holistic point of view. The hemispheres act locally when surgically separated, but function as a unit when linked together in the normal, unmodified brain. So the brain is both specialized and unified. But how is it organized? How does this marvelous and mysterious organ work?

The human brain remains the most complex structure in the known universe. And while more has been learned about it during the past two hundred years than in all previous history, we are still at a very early stage in our understanding.

For instance, what explains the complexity of the human brain? Size alone is not sufficient. An average human brain weighs 1,300 grams (about 3 pounds), roughly the same as that of a bottlenose dolphin (1,500 grams). In comparison, a sperm whale's brain weighs 7,800 grams, an elephant's 6,000 grams. All intelligent creatures, surely, but not as capable as humans. When it comes to surface area, the human brain is similarly outclassed by that of the African elephant and other large animals. So it is with the brain-to-body ratio and other parameters of size.

Blood flow to the brain gives us our first clue to what makes our brain unique. About 15 to 20 percent of the blood flow leaving the heart is destined for the brain. In order to function normally, the brain needs a reliable fund of oxygen, glucose, and other nutrients. If this supply is cut off for only 8 to 10 seconds, unconsciousness results. If oxygen deprivation lasts over 30 seconds, permanent damage may result.

It's estimated that the human brain contains 100 billion or more neurons and many times as many supporting cells. Its outer surface, the cerebral cortex, alone contains 30 billion neurons. But more important than the brain's total number of cells is

its total number of synapses, or neuronal connections—places where brain cells meet. The 30 billion neurons in the cerebral cortex establish 60 *trillion* synapses, an astounding average of about 2,000 synapses for each neuron! The combined circuitry that links together all the neurons of the cerebral cortex totals on the order of several hundred thousand miles. All this connectivity exists, incidentally, within a thickness of 1.5 to 4.5 mm.

But the brain may be even more complex than previously believed, for its neurons are outnumbered at least ten to one by its glial cells. What's more, these supporting cells (*glia* means glue in Greek) are now believed capable of encoding and transmitting information on their own. If this is true, then the number of elements involved in information transfer, along with their interactions, represents a truly inconceivable number, far in excess of the number of particles in the known universe.

————————

As an aid to dealing with such a level of complexity, it's helpful to keep in mind a few principles:

First, the brain operates simultaneously on several interacting levels, which range from processes observable by the naked eye to events invisible even to the world's most powerful microscopes.

Second, in order to understand the brain, it is necessary to get into the habit of simultaneously conceptualizing the brain's operation at these different levels. This book will help you do that.

Third, at each level of its operation the brain is concerned with one unifying process: the integration and processing of information. Just as there is information in the content of a sentence, so there is information in the propagation of a nerve impulse

and the union of neurotransmitters—the messenger chemicals that enable neurons to communicate with each other—and their receptors. Since this last operating principle is so important, let me illustrate it with a couple of examples.

As you read these words, photons of light bounce off this page, strike the retina of your eyes, and activate chemical processes that lead to the initiation of electrical impulses that pass along the optic nerves leading from the back of your eyes to your brain. These impulses are processed within several way stations along the path to your visual center, in the occipital cortex located at the back of your brain. Once there, the electrical signal is further processed via a vast network of intercommunicating nerve cells, which convert the electrical message into a chemical one.

A chemical messenger—neurotransmitter— is released by each sending neuron (known as the presynaptic neuron); it travels to a receiving neuron (the postsynaptic neuron). The action of the neurotransmitter and its receptor on the postsynaptic neuron results in the initiation of another electrical impulse that travels to the next neuron, and the electrical-chemical-electrical sequence is repeated.

Although we will explain all this in greater detail later, this simplified description captures the essence of the process: The message (consisting of words on a page) is translated into the "foreign languages" of electricity and chemistry.

Another example: If you were to attend one of my brain lectures, my words would be conveyed to you via waves created by the vibration of air produced by the movements of my lips and tongue. These waves would stimulate your tympanic membranes—eardrums—and result in the propagation

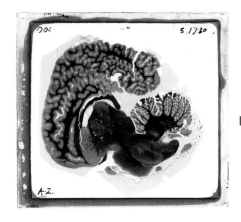

of an electrical impulse along the auditory nerves leading, eventually, to the auditory cortex. From here my words—now in the form of millions of interacting neuronal connections within the auditory association cortex and the rest of the brain— would be recognized not as electrochemical events but as words and meaning.

In both of these examples the brain transforms words and information into the harmony of nerve impulses, neurotransmitters, and receptors, only to re-transform them at higher levels of brain organization into the coinage of human language. This translation requires a decoding of information from one level to another. How is this accomplished? What is the relationship of a written or spoken word to an electrical impulse and later to a series of chemical reactions? In short, what is the "language" spoken within the brain and how can we learn to understand and speak it? The common element, of course, is information. Each level involves information existing in a different form and spanning levels ranging from words to ions.

In order to understand the brain it is helpful therefore to conceptualize its operation on four different levels. Although each of these levels will be described in greater detail in the next few chapters, here's an overview of the processes underlying the brain's operation.

• LEVEL ONE: The brain is a physical object that can be seen, held, and touched. It is about the size of a grapefruit, divided into two hemispheres by a midline cleft extending from front to back.

• LEVEL TWO: Events can no longer be seen directly but must be studied by means of special instruments. This includes such items as neural pathways. For example, a slight touch or a pressure to a fingertip results in the activation of special receptors in the finger that activate an electrical impulse that travels along a peripheral neuron from finger to the spinal cord. The impulse ascends the spinal cord to the thalamus, a special way station located beneath the cerebral hemispheres. After some modification within the thalamus, the impulse streaks to the sensory cortex of the cerebrum. Here the impulse is integrated within the network of ten billion neurons. Many of the connections are stimulatory, enhancing the impulse so that it stands out from the background. Other connections exert an inhibitory effect on surrounding neurons, thus sharpening and defining the sensation originating at the fingertip.

• LEVEL THREE: Events occur on a molecular scale. The nerve impulse (also known as the action potential) travels like a wave along the sending, or pre-synaptic neuron. As it reaches the end of that neuron, it activates the release of neurotransmitters that travel across the synapse, a gap separating the sending neuron from the receiving, or post-synaptic neuron. There, the neurotransmitters lock on to specialized receptors.

• LEVEL FOUR: Such neurotransmitter-receptor interactions result in the passage of ions (electrically charged atoms) into and out of the nerve cell. This movement activates cell processes responsible for every operation performed by the human brain.

Over the next several chapters we will flesh out the details of how events unfold on these four levels. First, let's explore the brain at level one—the "naked eye" level, which doesn't require the use of any specialized instruments.

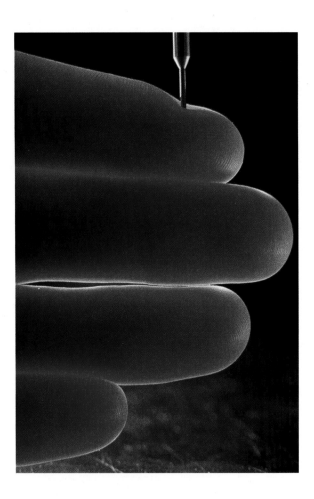

Skin

Axon

Myelin
sheath

Axon

■ Analysis of a touch: The pressure of a
probe on a fingertip (above) activates
thousands of touch receptors densely
clustered near the skin surface, triggering
an electrical discharge that pulses into
the long axon of a sensory nerve (right).
An insulating layer known as the myelin
sheath wraps the axon throughout its
length, helping it conduct electrical signals
from the skin to the spinal cord at up to
425 feet per second. From there, signals
are relayed to the brain stem, then to
the thalamus and the cerebrum's sensory
cortex—all in less than the blink of an eye.

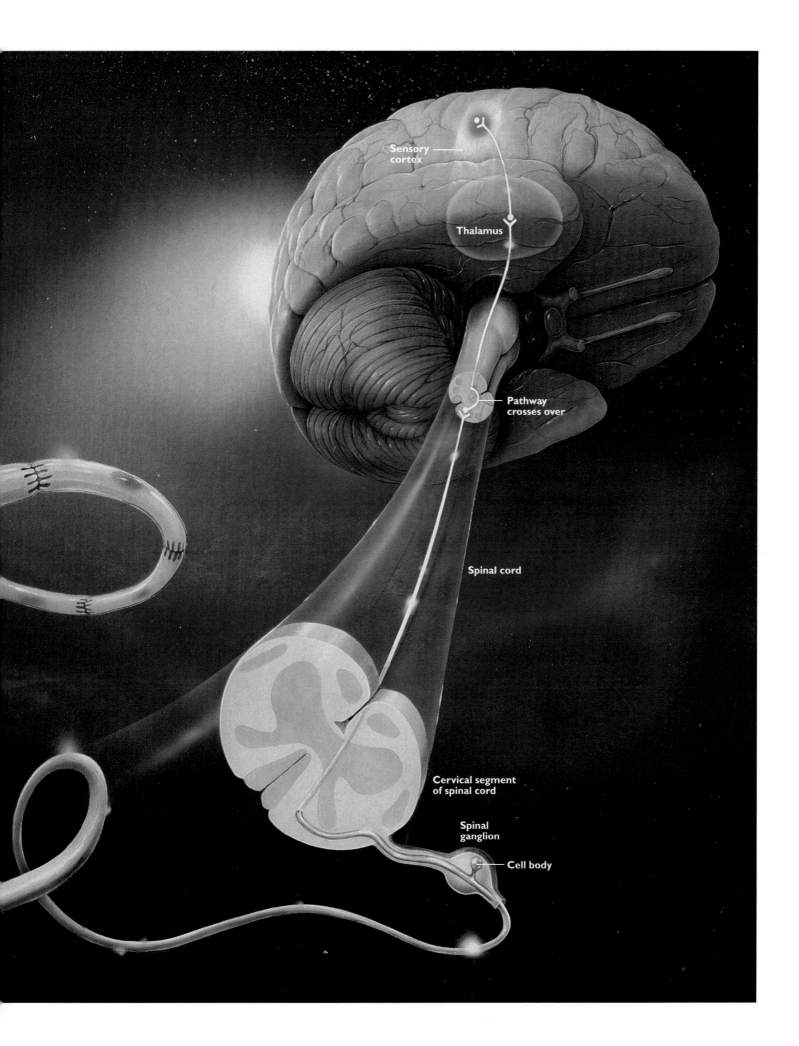

Sensory
cortex

Thalamus

Pathway
crosses over

Spinal cord

Cervical segment
of spinal cord

Spinal
ganglion

Cell body

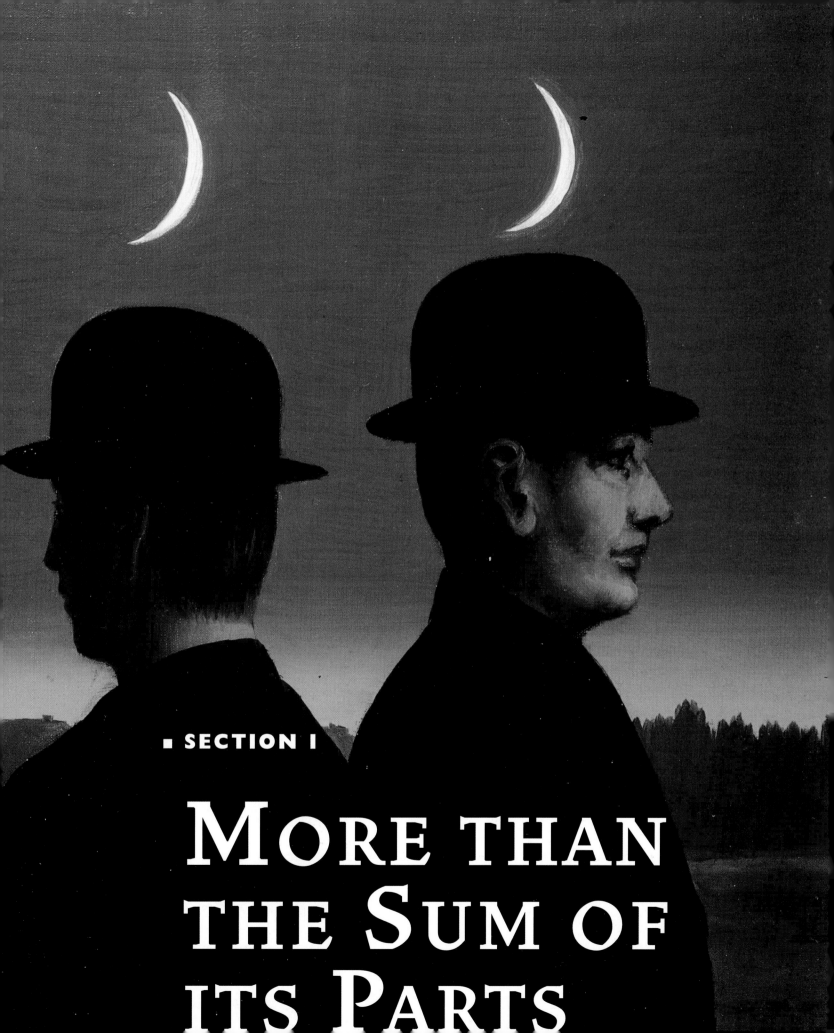

MORE THAN THE SUM OF ITS PARTS

Neuropathologist Barbara Crain discusses brain abnormalities in the autopsy room at Johns Hopkins University. While autopsy can show structural damage, it reveals nothing about how the brain functions electrically and chemically.

PREVIOUS PAGES: As if pondering three views of reality, Belgian surrealist René Magritte created *The Masterpiece or The Mysteries of the Horizon* in 1955. A prime exponent of magic realism, he often used familiar objects in an unfamiliar setting.

ARCHITECTURE OF THE BRAIN

Viewed from the side, each half of the brain resembles an old crumpled boxing glove. Most of what you see is called the cerebral cortex. This is the outermost part of the forebrain, which includes some 85 percent of all brain tissue. The front, middle, and back of each "glove" correspond to the frontal, parietal, and occipital lobes, while the thumb represents the temporal lobe.

An obvious feature of the cerebral cortex is its highly convoluted surface, which serves the practical purpose of increasing the surface area of this thin outer layer of the hemispheres, where the neurons (brain cells) are located, without a corresponding increase in volume and size.

To get a feeling for the advantage of a folded surface over a smooth one, place a fully opened handkerchief on a table. Notice that it takes up a square foot or so of surface area. Now crumple it and place it in your pocket. Its surface area remains the same—but it now fits in the relatively small volume of your pocket. A similar crumpling of your cerebral hemispheres enables a large surface area of brain to fit within your skull's limited capacity.

Such an arrangement not only provides increased brain area per unit of volume—a more complex brain in a small package—but also is indispensable to human survival. If all our functionally active brain tissue resided on a smooth surface instead of within folds and crevices, our heads would have to be enormously enlarged just to accommodate it all. Indeed, our skulls would be so large that they could not fit through the birth canals of our mothers. Even birth by Caesarean section wouldn't help much, for we all would have such "swelled heads" throughout our entire lives that

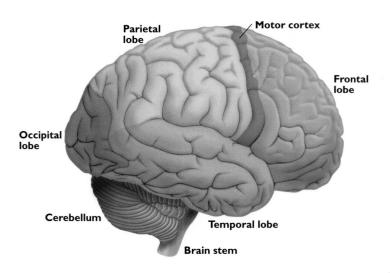

we'd be eternally topheavy, forever losing our balance and falling flat on our faces.

Let's look again at the cerebrum, and the four lobes that make up each hemisphere:

+ FRONTAL LOBE. Located toward the front of the brain, the frontal lobe contains almost 50 percent of the volume of each cerebral hemisphere. In addition to its importance in personality, emotion, and executive control (illustrated by the intriguing case of railroad foreman Phineas Gage), the frontal lobe plays important roles in language, movement, planning, and consciousness.

+ PARIETAL LOBE. Just behind the frontal lobe, the parietal lobe processes every sensation except smell, which connects directly to a more interior part of the brain, the limbic system. Think of the parietal lobe as a sensory integrator responsible for our sense of bodily position. It is separated from the frontal lobe by the narrow motor cortex, home of neurons that direct the body's motor nerves and thus bodily motion.

+ TEMPORAL LOBE. Thumb of the "boxing glove," the temporal lobe contains areas involved in hearing and understanding speech. It also contains connections to the hippocampus and amygdala, which are important in learning, memory, and emotion. In addition, the temporal lobe plays a prominent part in integrating our inner experiences and providing us with our sense of identity.

+ OCCIPITAL LOBE. Located at the back of the hemisphere, the occipital lobe contains regions important in visual perception and processing.

Each lobe is anatomically and functionally specialized. The density and arrangement of nerve cells in the optical cortex differs from what's in the auditory cortex, which differs from the cortical areas that concentrate on more evolved expressions of mind, such as thinking or remembering.

Of all the lobes, only the occipital is dedicated to straightforward processing of a single sensation, vision. The other three each dedicate a small portion, about 25 percent, to simple sensory or motor activities. The remaining 75 percent makes up what's called the association cortex—a vast network of communicating fibers that unifies our diverse perceptual and behavioral experiences. As a result of this network we don't see the world in terms of separate sights, sounds, and sensations. Instead, our brain unifies and synthesizes all our experiences into a whole, relying on those fibers of the association cortex.

The entire surface of both hemispheres, the cerebral cortex, consists of a thin rind—only about two millimeters thick—thinner than the cover of this book. Yet the cortex is full of neurons and supportive glial cells. Responsible for all our mental activity, it distinguishes us from all other creatures. It is also known as the gray matter, due to its dull appearance when the brain is cut in cross-section. Travelling to and from the gray matter are extensions of the neurons, referred to as nerve fibers, which connect to other areas of the brain. A fatty insulating substance, myelin, surrounds the nerve fibers, giving them a pale and milky appearance that explains their common name: white matter. Interspersed among the connecting nerve fibers like islands in an archipelago are isolated centers of gray matter we call the subcortical nuclei.

The human cerebral cortex is the most intricately organized and densely populated expanse of biological real estate in the world. One cubic millimeter of it—roughly the size of a coarse grain

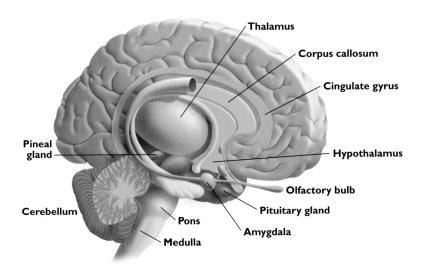

Thalamus

Corpus callosum

Cingulate gyrus

Pineal gland

Hypothalamus

Olfactory bulb

Cerebellum

Pituitary gland

Pons

Amygdala

Medulla

of sand—contains about 100,000 nerve cells. Since it's estimated that the average nerve cell, or neuron, receives input from about 100,000 others, the nerve cells contained within that single sand grain make about ten billion connections.

To explore the brain's interior regions, imagine separating the two hemispheres and viewing one of them from the inner side. In this view you can see the brain's three main divisions: the forebrain or cerebrum, the brain stem, and the cerebellum. Included among the forebrain structures are:

• The CORPUS CALLOSUM, an eight-million-nerve-fiber bridge that connects the two hemispheres and coordinates their activities.

• The CINGULATE GYRUS, which along with adjacent cortical matter is known as the limbic lobe. "Limbic" means border, and the limbic lobe borders the corpus callosum. It and other subcortical areas connected to it form the limbic system, which is involved in experiencing and expressing emotions.

• The THALAMUS, a way station for nerve impulses on their way from the periphery to the cerebral cortex and other parts of the brain. Taken from the Greek word for "inner room," the thalamus is located just outside the main entrance to the cerebral hemispheres.

• The HYPOTHALAMUS and PITUITARY, two small but crucial structures immediately beneath the thalamus. They are devoted to the maintenance and control of bodily processes such as temperature regulation, reproduction, and hormone production.

• The PINEAL GLAND, Descartes' so-called seat of the soul, now known as an important regulator of sleep and wakefulness.

Our internal side view also reveals the brain stem, the second of the brain's three major divisions. It is composed, from the top downward, of the midbrain, the pons, and the medulla.

The midbrain is the point of origin of the third and fourth cranial nerves, which move the eyes and control the size of the pupils. Together, the pons—Latin for "bridge"—and medulla contain the origins for the fifth through twelfth cranial nerves, which are responsible for sensations and movements of the face, outward movements of the eyes, also taste, hearing, swallowing, tongue movements, and other functions. Deep within the pons and stretched along its entire length resides a major portion of the reticular activating system. This complex and loosely arranged collection of cells extends the length of the brain stem; it helps keep us awake and alert.

The medulla, the lowermost portion of the brain stem, merges with the spinal cord and can be thought of as an extension of it. Within the medulla are nuclei—clusters of nerve cells responsible for hearing, balance, and head and neck movements. The medulla also contributes to the control of fundamental life processes such as blood pressure, heart rate, and breathing.

Finally, arched over the pons and the medulla and resting beneath the cerebrum's occipital lobes is the cerebellum. Long known to be important in the coordination of posture, movement, and balance, the cerebellum also plays a vital role in cognition: the processing of information.

In order to visualize the deeper structures hidden within each cerebral hemisphere it's helpful to slice the brain crosswise, like a loaf of bread, and look at

One of the most difficult parts of the body to study, the brain resides in a bony box, swathed by several layers of protective tissues. Recent developments in imaging, as well as vastly improved surgical techniques, have greatly improved our view of the brain's architecture and workings. We know that each cerebral hemisphere, right and left, controls the opposite side of the body. Different areas within each cerebral hemisphere specialize in certain functions—the motor cortex, for example, helps control conscious movement. Other structures apart from the cerebrum, such as the hippocampus, which aids memory, have their own jobs as well. Neurons, the networking wonders of the brain, send and receive electrochemical communications in mere thousandths of a second at connection points called synapses.

Synapse

Where neurons meet: Specialized arms called dendrites extend from one nerve cell to the next, nearly touching in a tiny space called a synapse (right). Here, chemical neurotransmitters released by one neuron cross the gap and stimulate the receiving neuron to send a chemical message to the next cell in line.

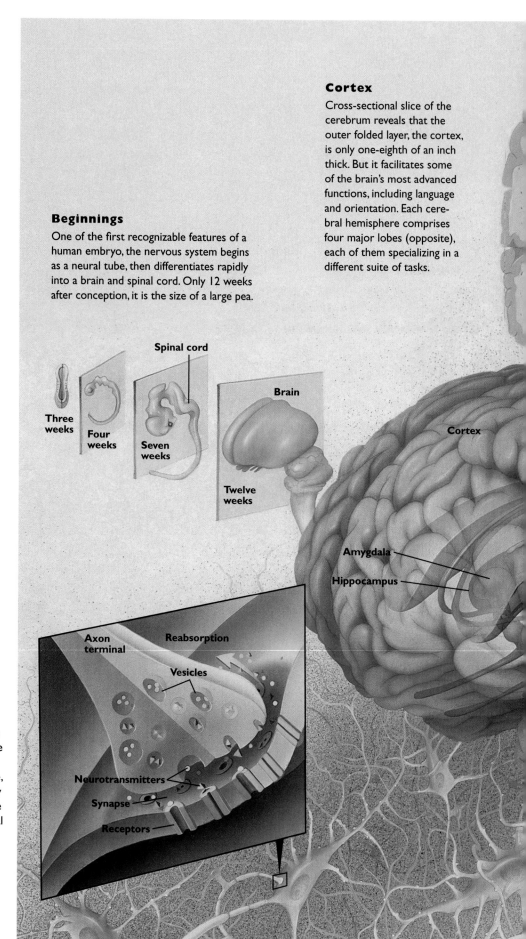

Beginnings

One of the first recognizable features of a human embryo, the nervous system begins as a neural tube, then differentiates rapidly into a brain and spinal cord. Only 12 weeks after conception, it is the size of a large pea.

Cortex

Cross-sectional slice of the cerebrum reveals that the outer folded layer, the cortex, is only one-eighth of an inch thick. But it facilitates some of the brain's most advanced functions, including language and orientation. Each cerebral hemisphere comprises four major lobes (opposite), each of them specializing in a different suite of tasks.

Spinal cord

Three weeks

Four weeks

Seven weeks

Brain

Twelve weeks

Cortex

Amygdala

Hippocampus

Axon terminal

Reabsorption

Vesicles

Neurotransmitters

Synapse

Receptors

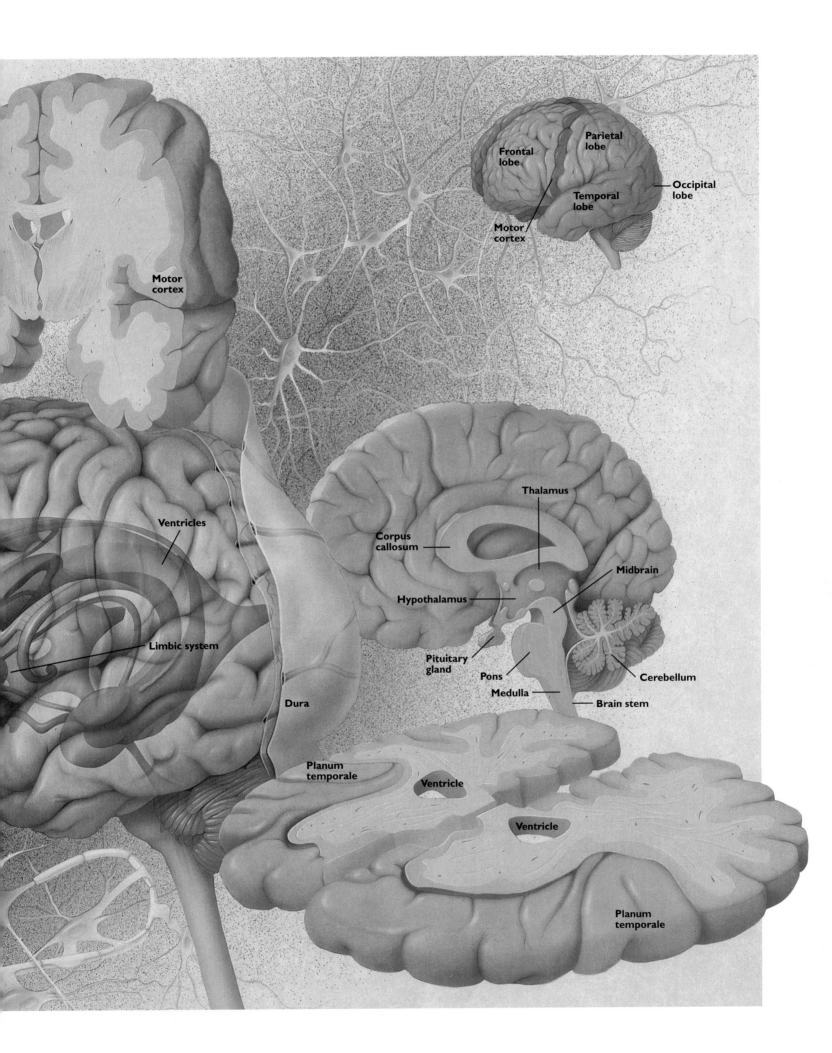

Motor cortex

Frontal lobe

Parietal lobe

Temporal lobe

Occipital lobe

Motor cortex

Ventricles

Limbic system

Dura

Planum temporale

Thalamus

Corpus callosum

Hypothalamus

Midbrain

Pituitary gland

Pons

Medulla

Cerebellum

Brain stem

Ventricle

Ventricle

Planum temporale

one of the center slices. The butterfly-shaped structure you see in the middle comprises the two lateral ventricles. In the immediate vicinity are the basal ganglia: large nuclei (islands of gray matter) composed of three main substructures: the caudate and the putamen—together called the striatum, because of their striated appearance—and the globus pallidus.

The basal ganglia organize and coordinate movements that do not require conscious processing. Often they take over after the conscious aspect of learning is completed. For instance, when you first learn the tango, you have to concentrate and focus your efforts to master the steps. At this stage in your learning, the motor centers in the cerebral hemispheres play the dominant role, while you consciously attend to your movements. But after more instruction and practice you eventually become so skilled that the steps become almost second nature, like driving a car or riding a bicycle. At that point, the emphasis has shifted from the cerebral cortex to the basal ganglia.

Immediately below the basal ganglia are the forebrain nuclei. Comparatively small in humans, these nuclei exert an influence on mental function far in excess of their diminutive size. Abnormal structure and function of one of them, the basal forebrain nucleus of Meynert, is associated with the devastation wrought by Alzheimer's disease.

Another small but powerful structure buried deep in the hemispheres is the amygdala. It lies in front of the hippocampus, toward the front of the temporal lobe. Part of the limbic system, it is an important participant in the experience and expression of emotion; it helps evaluate an event's emotional significance. Hearing footsteps on the stairs at midnight when you're alone, for example, activates the amygdala and generates anxiety. But no anxiety is experienced when you hear those same sounds on a night when friends are staying over. That's because the amygdala hasn't signaled the presence of a potentially dangerous situation.

The amygdala is also important in memory and works in tandem with the nearby hippocampus, which serves as entry point for the formation of memories. Sights, sounds, and other sensory information enter the hippocampus, where they are incorporated into a memory for that specific moment. Damage to the hippocampus on both sides of the brain interferes with the brain's ability to form and recall new memories.

At this point we have identified the major brain areas and structures discussed throughout this book. Many are small; many are located deep within the brain, such as the ventricular system, a series of fluid-filled and interconnected spaces that extends from the cerebrum into the brain stem. Residing within the ventricles is the choroid plexus, which contains clusters of specialized secretory cells that produce the cerebrospinal fluid that bathes the brain and provides a soft protective cushion.

Additional protection for the brain is provided by three membrane layers, or meninges, found just beneath the bony skull. Innermost is the pia mater (Latin for "tender mother"), a thin and delicate layer of cells that is intimately attached to the brain's surface. Next is the arachnoid, so called because it resembles a spider's web. Between it and the pia mater flows the cerebrospinal fluid. Then comes the stiff and unyielding outermost layer, the dura mater

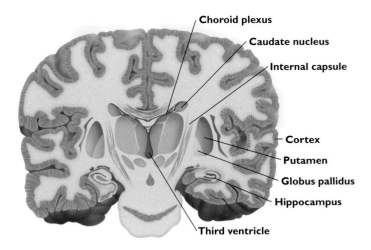

Choroid plexus
Caudate nucleus
Internal capsule
Cortex
Putamen
Globus pallidus
Hippocampus
Third ventricle

(hard mother), which is partly attached to the skull. Head injuries sometimes cause bleeding that seeps between the dura mater and the skull, resulting in a potentially fatal subdural hemorrhage.

But of course anatomy is only part of the brain's story. How is it organized? First, although the brain influences every organ in the body, it is physically isolated and protected by the skull, the meninges, and the ventricles with their enclosed cerebrospinal fluid.

Second, the brain operates both locally and holistically. While it's true that certain areas are specialized for specific purposes, this organ can only be understood as one highly complex and integrated functional unit. Indeed, the great majority of its neuronal connections involve "cross talk" among neurons rather than the transfer of information to and from the rest of the body. The large association cortex is responsible for this information flow from one part of the brain to another.

Third, despite the brain's billions of neurons and their trillions of connections, no nerve cell is very remote from another. Remember the play *Six Degrees of Separation*, which made the point that any two people on Earth can be connected by no more than six linkages? A similar situation exists in regard to nerve cell connections.

Fourth, size and location are not always paramount in determining the importance of a brain structure. Without the medulla, all heart and lung activity would cease; yet this part of the brain stem is not much bigger than a fingertip. The hypothalamus constitutes less than one percent of the total volume of the brain, despite its role in regulating temperature, food and water intake, and other bodily processes. Pioneering neurosurgeon Harvey

Cushing, first to recognize the importance of the hypothalamus, commented on the discrepancy between its size and its importance: "Here in this well concealed spot, almost to be covered with a thumbnail, lies the very mainspring of primitive existence—vegetative, emotional, reproductive— on which, with more or less success, man has come to impose a cortex of inhibitions."

Fifth, the brain exhibits hierarchical organization. High-level, complex, predominately symbolic activities such as thinking and consciousness evolved more recently than lower level, more automatic functions. Just as walking and gesturing preceded speech and analytical geometry, so too the basal ganglia and primitive cerebral cortex preceded the flowering of the frontal lobes. As a consequence of this sequence, injury to those areas of the brain that have evolved more recently often results in the expression of lower level operations. For instance, a stroke patient may not only lose the powers of speech but may also revert to primitive reflexes ordinarily observed only in infants. A similar process may occur behaviorally, as Sigmund Freud appreciated when formulating his concept of regression, the idea that the "id" can erupt and overcome the "ego." Freud compared the situation to "a man on horseback who has to hold in check the superior strength of the passion-driven horse."

Finally, the brain exhibits what neuroscientists refer to as plasticity. Each and every experience over a lifetime changes the brain in some way. The changes may not be obvious or even demonstrable, but they occur nevertheless. Since the details of this occur at the microscopic level, we'll take up plasticity in more detail in the next chapter, where we explore *level two*, the brain's electrochemical activity.

ON THE CELLULAR LEVEL

Humorist Robert Fulghum reached the following conclusion about the brain: "Now I can kind of understand the mechanical work of the brain—stimulating breathing, moving blood, directing protein traffic. It's all chemistry and electricity. A motor. I know about motors.

"But this three-pound raw-meat motor also contains all the limericks I know, a recipe for how to cook a turkey, the remembered smell of my junior-high locker room, all my sorrows, the ability to double-clutch a pickup truck, the face of my wife when she was young, formulas like $E=mc^2$ and $a^2+b^2=c^2$, the Prologue to Chaucer's *Canterbury Tales*, the sound of the first cry of my newborn son, the cure for hiccups, the words to the fight song of St Olaf's [sic] College, fifty years worth of dreams, how to tie my shoes, the taste of cod-liver oil, an image of Van Gogh's 'Sunflowers,' and a working understanding of the Dewey Decimal System. It's all there in the MEAT."

As we've seen, the "meat" of the human brain consists of somewhere in the range of 100 billion neurons—nerve cells—and several times as many glial, or supporting, cells. In most ways neurons structurally resemble other body cells. The neuron's uniqueness stems from its widely diverse and often aesthetically pleasing geometries. But despite the great variation in neuronal shape, each neuron is engaged in transmitting information and therefore shares several common features.

First, it has a cell body, containing the nucleus and cytoplasm common to all cells. It also has dendrites, branches that serve like antennae; each receives information from other neurons in the form of electrical impulses.

The number and complexity of dendrites varies enormously from cell to cell. Some neurons seem to have none, while others possess dendritic configurations reminiscent of a densely branched tree. In general, the more dendrites a neuron has, the greater the number and complexity of its connections. A neuron with elaborate and extensive dendritic patterns can communicate, directly or indirectly, with as many as 100,000 other neurons.

However many its connections, each neuron keeps a careful tally of the impulses its dendrites receive. In response to these impulses it may fire a signal to other neurons via its relatively long axon. Each neuron possesses only one axon, which can vary immensely in length, from a few millimeters to over a meter. Unlike dendrites, which receive messages, the axon transmits them, passing information down its entire length, from the neuron's cell body to the next neuron, by means of an action potential—a self-generating wave of electricity.

The axon's tip marks the synapse, a tiny gap between neighboring neurons. While conduction along the axon is entirely electrical, the synapse involves chemical transmission. Neuroscientists arrived at this conclusion via an astute observation by German physiologist Hermann von Helmholtz.

In one experiment, Helmholtz measured the time required for a nerve impulse to travel the length of an axon. By carrying out the experiment under differing conditions, Helmholtz made an important observation: Lowering the temperature of the nerve significantly reduced conduction velocity. Since

■ Constant interplay of impulses occurs among the myriad neurons of the sensory cortex. Here, squat pyramidal cells receive incoming signals, while darker basket cells act as inhibitors, preventing overload of the sensory circuitry.

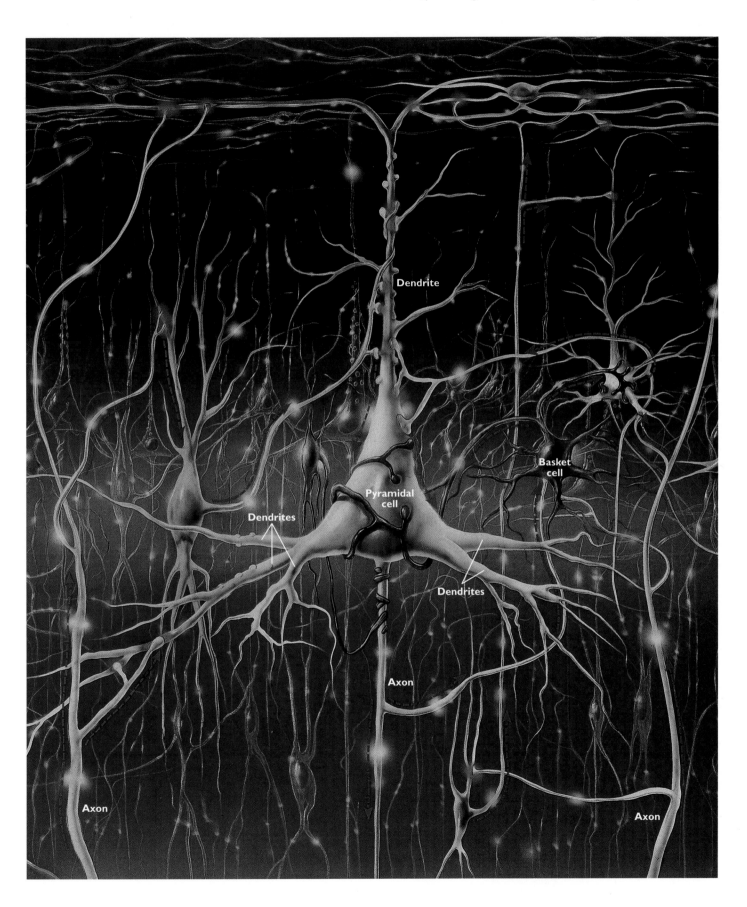

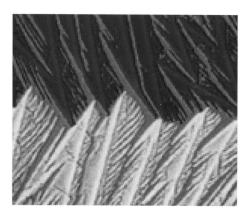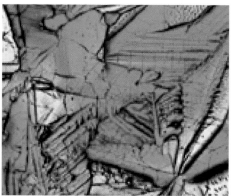

■ Chemicals of thought and mood: Crystallized and photomicrographed in polarized light, the important neurotransmitters acetylcholine (right) and serotonin (far right) display different structures, reflecting their very different chemical properties.

temperature variations strongly influence chemical reactions but exert little effect on electrical conduction, Helmholtz concluded that chemicals somehow played a role in nerve conduction. But if the nerve impulse was partly chemical, what was that chemical and how might it be isolated? The answer to this intriguing question, believe it or not, appeared to the physiologist Otto Loewi in a dream in 1920.

"The night before Easter Sunday…I awoke, turned on the light, and jotted down a few notes on a tiny slip of paper," Loewi later recalled. "Then I fell asleep again. It occurred to me at six o'clock in the morning that during the night I had written down something most important, but I was unable to decipher the scrawl. The next night at three o'clock the idea returned. It was the design of an experiment to determine whether or not the hypothesis of chemical transmission was correct. I got up immediately, went to the laboratory, and performed a simple experiment on a frog heart according to the nocturnal design."

Loewi's "simple experiment" involved suspending two frog hearts along with their attached nerves in a single container of solution. He then electrically stimulated one of the nerves. Within seconds, the heart attached to that nerve beat more slowly. But then the second heart slowed as well! Since direct electrical stimulation had not been applied to that heart's nerve, Loewi concluded that some chemical must have been released by the first nerve into the solution, and it had slowed down the second heart. Today scientists identify that chemical as acetylcholine, one of a number of neurotransmitters that occur both in the body's nerves and in the brain.

Thus communication within the brain is not purely electrical or chemical but electrochemical. Indeed, the neuron is best thought of as a specialized cell that has evolved for the transmission of electrochemical messages.

The electrical component of transmission—known as axonal conduction, since the outgoing impulse travels the length of an axon—demands some adaptations within the brain. Axons are not very good conductors of electricity, and in fact are far less efficient than a typical electrical wire. Several biological adaptations have evolved that improve the axon's poor electrical conductance.

For one thing, axons are wrapped in an insulating sheath of myelin, the same material that makes the brain's white matter appear white. Myelin speeds up the conduction of the axon potentials.

Secondly, the neuron's outer membrane is selectively permeable to different ions, principally sodium, potassium, chloride, and calcium. This selective ionic permeability is based on the presence of pores (known as ion channels) that permit certain ions to pass from one side of the membrane to the other. Thanks to variations in permeability at different stages of the nerve impulse, the concentrations of sodium, potassium, and chloride vary within and without the nerve cell. At rest, the nerve cell contains little sodium; most sodium remains outside the cell. The distribution of potassium is just the opposite: the lion's share is inside the cell.

Such concentration differences or "gradients" are responsible for the electrical potential that exists across the nerve-cell membrane. Changes in this membrane potential result in the propagation of an action potential. It is one of two methods for conveying information from one place to another, both within the brain and throughout the rest of the nervous system.

The second means of information transfer starts after the action potential has sped along the entire length of the axon and reached the end. Here, at what's called the presynaptic membrane, are gathered thousands of spherical structures—the synaptic vesicles—which contain one or more of the chemical messengers we call neurotransmitters.

When the action potential reaches the end of the axon, it causes the calcium channels in the nearby membrane to open, resulting in a massive rush of calcium into the nerve cell. This increase in intracellular calcium exerts a pinball effect on many components and processes.

First, the synaptic vesicles fuse with the plasma membrane, dumping their contents—the neuro-transmitters—into the synapse. These chemicals then wend their way across the synapse, traveling from the presynaptic nerve to the postsynaptic one, and attaching to specific receptors on the membrane of the postsynaptic neuron.

The binding of neurotransmitters to these receptors opens or closes certain ion channels on the membrane, creating a flow of ions that in turn generates a small electrical current, thus transform-ing the message back from chemical signals into electrical ones. This current changes the membrane potential of the postsynaptic neuron, making it more likely to fire an action potential down its axon and by this means forward information to the neuron next in line.

Originally, neuroscientists compared the fit of a neurotransmitter and its receptor to a key and its lock. Acetylcholine, for instance, would seek out its specific receptor and bind to it.

Actually, this analogy is oversimplified and too mechanical. For one thing, receptors are biological structures that can change their structure and identity. They can become more or less active, can increase or decrease in number, and can even change their identity: a serotonin receptor can become a norepinephrine receptor, and vice versa.

In essence, communication within the brain involves multiple neurotransmitters—acetylcholine, norepinephrine, serotonin, and perhaps many hundreds more—operating via multiple variations of receptors. Serotonin, for instance, has more than 15 subtypes of receptors. One practical consequence of this diversity is that specific drugs can be used to influence a single receptor subtype.

For example, several currently used drugs for depression and migraine headache affect some serotonin subtypes while exerting no influence on others. In addition, the brain can alter the number and composition of its receptors. Indeed, this is one of the mechanisms of action of the newer serotonin-based antidepressants we will discuss in more detail in a later section.

———————

Return for a moment to the synapse, to the binding of a neurotransmitter to its receptor and the result-ant opening or closing of ionic channels. This process can occur in one of two ways. Either the channel opens immediately (such receptors are referred to as ionotropic, since ions are involved) or the channel opens after a delay (these are called metabotropic receptors, since metabolic reactions must precede the movement of ions). The delay

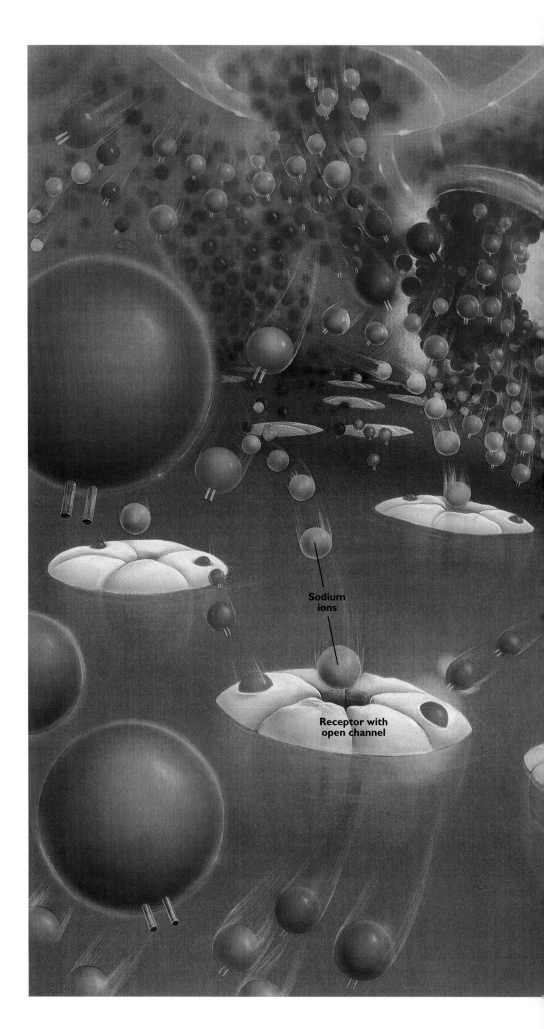

Sodium
ions

Receptor with
open channel

■ Molecule's-eye view of a synapse: As an impulse travels down a nerve to the synaptic bulb, thousands of vesicles spill their neurotransmitters into the synaptic gap, only a billionth of an inch wide. Each molecule of neurotransmitter binds with a suitable receptor on the target neuron, an act which prompts sodium ions to rush in and potassium ions to leave. This flow of ions excites the target nerve, thus generating an electrical impulse in that cell.

Synaptic vesicle

Target cell

Receptor with closed channel

Neurotransmitter molecules

Potassium ions

> **Emotion is always in the loop of reason.**
> —Antonio Damasio

results from the activation of intermediary molecules called G-proteins.

Built for speed, ionotropic receptors respond within a millisecond or two; their G-protein counterparts are designed for much slower and more prolonged responses, which may last minutes, hours, or even longer. Thanks to these two receptor families, the brain can store information along a continuum from milliseconds to years.

Although receptor research is comparatively recent, its practical applications were intuitively appreciated hundreds if not thousands of years ago. Whenever South American Indians rub curare on arrowheads, they are taking advantage of that poison's ability to block one class of acetylcholine receptors. Curare acts by preventing acetylcholine from occupying and stimulating receptors at the neuromuscular junction, where nerve and muscle meet. So, although the action potential can travel unimpaired down the nerve, the nerve cannot stimulate the muscle, resulting in paralysis. Death follows when the muscles involved with breathing become paralyzed.

A similar effect results from the venomous bite of a snake known as the banded krait. This snake's venom contains a neurotoxin called alpha bungarotoxin, which paralyzes prey (and enemies) by binding to a subtype of acetylcholine receptor and thus preventing that neurotransmitter from opening the ion channels on the membrane of the postsynaptic cell. As with curare, this results in muscular paralysis. Once again, the muscles involved in breathing can no longer be activated by motor neurons. Death by suffocation results.

Strychnine poisoning—a favorite method in Victorian novels for dispatching the unwanted—

blocks transmission at those synapses within the brain that use the inhibitory neurotransmitter glycine. Deprived of the normally inhibitory action of glycine, the brain becomes not paralyzed but overstimulated—and the victim dies as a result of uncontrollable seizures.

———————————

Today, man-made drugs rather than natural substances have become a primary chemical influence on brain function. Included here are those drugs synthesized by reputable pharmaceutical companies as well as the mind-bending agents cooked up in garages and basements by rogue chemists. Whatever their origin, drugs affect the mind and brain in various ways, which depend on the brain's own chemical composition.

Psychopharmacology is the study of how drugs affect the brain and thereby influence thought and behavior. Not every drug that influences neurons results in behavioral changes, of course. Remember your last trip to the dentist for a procedure requiring novocaine? That shot of novocaine deadened pain by altering the membranes on nerves leading from your teeth. Novocaine is a local anesthetic, so named because it induces localized nerves not to transmit painful impulses, and thereby indirectly alters your pain perception. But since it doesn't enter your brain, it has no effect on your mental state or behavior. (Of course, you may feel relieved at the absence of pain, but that isn't due to any direct effect on your brain.)

For a drug to alter the mind, it must enter the brain and somehow influence the chemistry of synaptic transmissions. In most instances, this involves changes in the receptors and neurotrans-

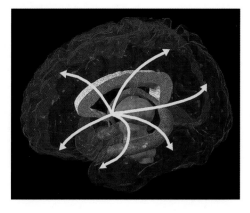

Acetylcholine—a key neurotransmitter involved with memory as well as one's emotional state—projects up to various areas of the cerebral cortex from the nucleus basalis and the limbic system.

mitters. According to current estimates there are more than 100 different chemicals that may act as neurotransmitters or neuromodulators (smaller chemicals that subtly influence the action of the neurotransmitters themselves).

Despite this plethora of chemicals affecting the brain, neuroscientists have concentrated on only a few of them. Thus it's possible—some would even say likely—that future discoveries concerning some less studied chemicals may provide important new insights into how the brain functions.

———————

For now, psychopharmacology is chiefly taken up with four neurotransmitters: serotonin, dopamine, norepineprine, and acetylcholine. Each occurs in the brain stem and extends upward to the limbic system, basal ganglia, and cortex. In addition, the brain contains small amino-acid neurotransmitters such as glutamate—the major excitatory neuro-transmitter—and GABA (an acronym for gamma aminobutyric acid), the body's major inhibitory neurotransmitter.

Until recently, neuroscientists believed that a given neuron produced only a single neurotrans-mitter. We now know that many neurons contain and release two or more neurotransmitters. This increases a neuron's versatility, modulating or enhancing its responses. Adding a co-transmitter is like replacing a simple off-on light switch with a rheostat, or dimmer.

While naturally occurring neurotransmitters act on specific receptors, man-made drugs have additional options. Some act like the brain's natural neurotransmitters and stimulate the corresponding receptors. They are called agonists. Others block the actions of a naturally occurring neurotransmitter, and are referred to as antagonists. Psychoactive drugs work by occupying a particular neurotrans-mitter's receptor. Even commonly used substances such as cigarettes, alcohol, tea, coffee, and most sodas contain chemical compounds that affect the functioning of your brain and therefore can be considered to be mind-altering drugs.

Thus we see that the brain, like the nerves of the body, is an electrochemical organ that is modifi-able by the environment in which it finds itself. Functionally it is composed of networks of neurons that are unique for every person in the world. Even the brains of identical twins are not the same, because they do not undergo the same experiences. These different experiences are embedded within the brain over time, in the form of distinct neuronal networks. Operationally, the brain employs these networks to transmit information via a two-step process involving the action potential (nerve impulse) and the chemical processes that occur among neuro-transmitters and their receptors.

Some neurotransmitters are excitatory; they activate additional neurons and thus increase the spread of activation potentials throughout the brain. Other neurotransmitters are inhibitory; attachment to their receptors decreases the likelihood of further neuronal stimulation.

Our understanding of the mysteries of the mind will continue to expand in tandem with new and wide-ranging scientific techniques for studying and imaging the living brain. Indeed the advances of the past two decades would not have been possible without vital contributions from physics, chemistry, and computer science. The next chapter explores our current abilities to see the brain at work.

WINDOWS ON THE MIND

Most early knowledge about the brain was based on a crude but serviceable blend of observation and deduction. Doctors would carefully note and catalog damaged areas of the brain while carrying out autopsies. They then correlated those injured areas with the symptoms and signs reported prior to the patient's death. Over the years this rough and ready method led to the compilation of brain "atlases" that linked areas of brain damage with a vast catalog of difficulties reported by patients suffering from various brain diseases. While such indirect methods proved useful and even led to the eventual emergence of neurology and neurosurgery, they didn't provide a means of directly observing the living brain.

The first painless and safe technique for making images of a live brain dates from the early 1970s with the development of x-ray computed tomography, or CT. CT uses a moveable x-ray tube that revolves around the patient's head while the head lies cradled within a doughnut-like configuration of sensitive x-ray detectors. The x-ray beam rotates from detector to detector and a computer converts the collected information into a series of images. Since the x-ray beam performs a complete revolution around the head, it gathers information from the entire brain. Visual "slices" of the brain (*tomos* is Greek for "section") can then be made through various planes and the corresponding images generated. CT served as a stimulus for the development of other imaging techniques.

One, magnetic resonance imaging, or MRI, relies on the tendency of atoms to behave like tiny compass needles when exposed to a magnetic field. The patient's head is placed within a huge magnetic

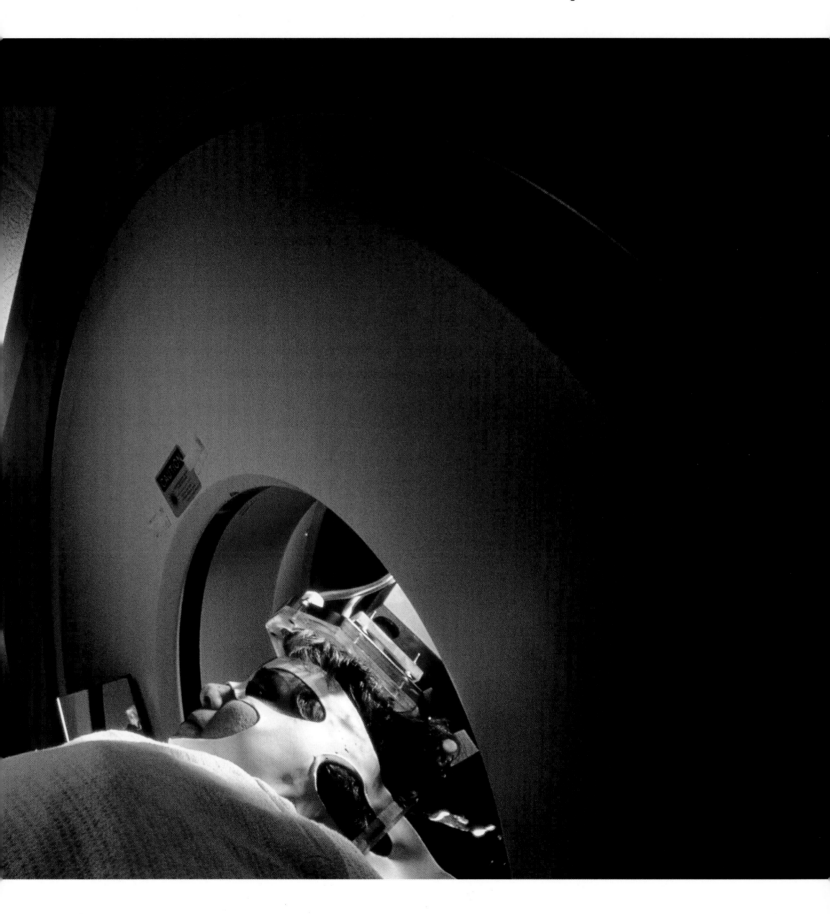

Reading from a monitor and then thinking of synonyms for certain displayed words, a volunteer in a PET (positron emission tomography) experiment helps scientists determine which brain areas become active during different mental tasks.

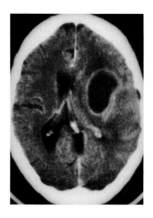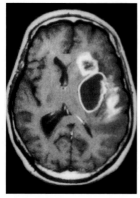

CT (far left) and MRI (left) scans of a glioblastoma multiforme, among the most lethal of all brain tumors. Like ink spilled on a carpet, these tumor cells infiltrate normal brain tissue without creating a distinct margin. Treatment usually involves surgery combined with radiation therapy or chemotherapy.

coil inside the MRI scanner. Under the influence of the magnetic field, the atoms within his brain tissue line up in a stable, ordered configuration. A pulsed radio wave is then turned on, which knocks the atoms out of alignment. Then the pulse is abruptly shut off, allowing the atoms to return to the original alignment, as imposed by the magnetic field. In the process of re-aligning, the atoms emit detectable radio signals that can be gathered, fed into a computer, and transformed into detailed brain images.

Sounds complicated, but it's very similar to what would happen if you flicked a compass needle with your finger. Normally, the needle tends to align with the direction of the north magnetic pole. Flicking the needle temporarily disturbs that alignment, but Earth's magnetic field soon returns the needle to its previous north-south orientation. During an MRI study, the atomic nuclei within your brain tissue act like so many compasses and, after their initial deflection by the pulsed radio wave, return to the position imposed by the surrounding magnetic coil.

MRI provides greater resolution than CT (less than a millimeter, compared to several millimeters for CT). But while both techniques can provide marvelously detailed information about brain structure, they don't evaluate function very well. They're rather like looking at the seating arrangement of a concert hall: They may give you an idea of where you'd like to sit, but they don't enable you to hear the music.

A variant of MRI called functional MRI, or fMRI, gives you the equivalent of not only the music but also the orchestral score. This technology is based on the fact that the amount of oxygen carried in the blood alters that fluid's magnetic properties. Specifically, hemoglobin emits different magnetic signals, depending on whether it is oxygen-enriched or oxygen-depleted. Functional MRI measures these differences.

Just like muscle and other body cells, the brain's neurons require the delivery of greater amounts of oxygen as they become increasingly active. Thus, active areas of the brain need more oxygenated hemoglobin; small nearby blood vessels dilate, increasing the flow of oxygen-rich hemoglobin to those brain areas. Functional MRI records and translates these dynamic shifts in the balance of hemoglobin's oxygen-rich and oxygen-poor forms into computer images that at any given time provide a window on ongoing brain activity. For example, if you're listening to music on your stereo while you're reading these sentences, an fMRI of your brain would show activity in brain areas that would include the occipital cortex (reading), the frontal cortex (thinking), and the temporal cortex (which contains music-associated brain areas).

An even more elaborate and expensive imaging technique, positron emission tomography (PET) relies on the radioactive decay pattern of electrically charged positrons. First a positron-emitting radioisotope such as carbon, nitrogen, oxygen, or fluorine is injected into a vein in the arm. The isotope eventually travels to the brain, where its positrons are emitted, only to collide less than a millimeter later with electrons. This collision generates gamma rays, which are easily detectable. Computers then construct images—tomograms—based on this decay sequence of the isotope. Because the most active areas of the brain need the most glucose and oxygen and, therefore, the greatest

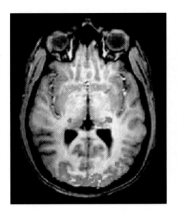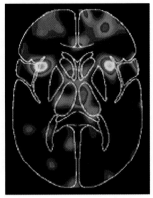

blood flow, they concentrate the greatest amounts of isotope—and show up most prominently in PET images. The introduction of PET enabled neuroscientists to localize specific mental operations. They found, for instance, that different brain areas "light up" on a PET scan, depending on whether the person views words displayed on a television monitor, hears those words, speaks them, or thinks of new words.

Both PET and fMRI provide measures of functional brain activity—that is, the actual music rather than just the seating arrangement. But their information is highly selective, providing only three measures of the living brain's metabolic changes: blood flow, oxygen use, and glucose use. Neither PET nor fMRI provides any information about electrical activity, nor do they operate rapidly enough to capture the fluctuations that underlie normal brain activity. Neurons communicate in millisecond intervals—but it takes about 40 seconds to obtain the data required for a single PET image of blood flow in the brain. Functional MRI measurements are faster, but they still aren't up to the speed of moment-to-moment neuronal processing. To monitor events in the millisecond range, neuroscientists depend on electrical recording devices.

An electroencephalogram, or EEG, is a recording device that registers the brain's ongoing electrical activity and translates that activity into a series of wiggles recorded on a long roll of paper. EEGs are most useful in the detection of epilepsy. The abnormal electrical discharges of the epileptic brain show up as abnormal waves that help the neurologist arrive at a diagnosis and plan of drug treatment.

On occasion, the EEG also can provide surprising explanations for seemingly inexplicable behavior. For instance, a person who falls asleep during an important meeting may later be diagnosed by EEG testing to suffer from narcolepsy, a rare disorder marked by episodes of uncontrollable sleepiness.

While EEGs can be helpful in the diagnosis of epilepsy and other disorders, they provide only limited useful information about normal states of consciousness. In addition, the EEG represents a general readout of overall brain activity; it is too imprecise to monitor localized activity in a specific brain area. Like a tape recording of a large banquet, it preserves the buzz of conversation but can't track the specific discussion of two people sitting at a particular table.

In order to home in on the conversation at a particular brain site, the EEG must be linked with a computer that can track the brain's response to a specific stimulus, such as a light of certain color, or a sound of certain pitch. The stimulus is repeated many times; occasionally a different stimulus is inserted, "surprising" the brain. The computer averages out any electrical activity that is not related to the surprise stimulus, thus discounting the "noise" of the brain's ongoing background electrical signals.

While computer averaging of EEG signals provides more specific information, it still leaves a lot to be desired in terms of precisely locating ongoing electrical events within the brain. Returning to the banquet-hall metaphor, computerized EEG directs our attention toward a general area of the room, not to the specific table and conversation we're interested in. To accomplish that we have to

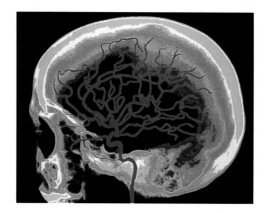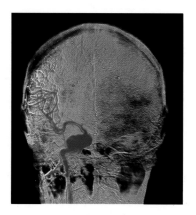

use a new and revolutionary instrument called the magnetoencephalograph, or MEG.

Magnetoencephalography records the magnetic fields produced by the living brain. These fields are extremely small, nearly a billion-fold smaller than the ambient magnetic fields produced by the Earth and many other sources of magnetism in our environment. One consequence of MEG's extreme sensitivity is that this technology is very expensive, requiring a magnetically shielded room usually composed of layers of aluminum and a special iron-nickel alloy.

But MEG can locate magnetic field changes to the nearest cubic millimeter of cerebral cortex. It can also time any changes taking place within the brain to the nearest thousandth of a second. With MEG, you can finally zero in on the specific table and hear every word of the conversation.

Each of these different techniques—CT, MRI, fMRI, EEG, and MEG—provides a unique window on brain structure or function. Each has strengths and weaknesses. In general, CT, MRI, fMRI, and PET are good at achieving excellent spatial resolution but do less well at time resolution. EEG can pin down the timing of events very nicely but doesn't localize events very well. MEG does both—but it's expensive and, so far, underutilized.

At the moment, each imaging technique is used for specific indications. If a neurologist suspects that a patient's sudden headache may be the result of a brain hemorrhage, he will order a CT because it is more sensitive than MRI to the presence of blood resulting from a rupture of one of the brain's major blood vessels. If blood is found on the CT scan, MRI can then be used to help the neurosurgeon locate the rupture.

In a case involving multiple sclerosis, in contrast, the neurologist will likely select MRI because of its ability to show fine details of the white matter, which is selectively attacked in MS.

In each instance, the goal is to choose the imaging technique that provides the optimal window on a selected area of brain functioning.

————————————

Will further developments in imaging technologies dramatically extend their individual usefulness? This seems unlikely. The time resolution of fMRI, for instance, will remain low compared to EEG because changes in the oxygenation of cerebral blood occur more slowly than the electrical and chemical events that transpire within the neuron, a time frame measured in milliseconds.

Since the brain functions at such different temporal and spatial scales, it's likely that neuroscientists will continue to speak in terms of plural windows on the brain rather than a single, all-encompassing picture window.

Now that we have examined the structure and function of the brain, along with the techniques available to plumb its mysteries, let's explore how this knowledge aids in our attempts to understand the mysteries of the mind.

■ Lifeline of a living human brain, the blood
vessels revealed in this angiogram serve
the most complex of organs, nourishing
the billions of brain cells that mysteriously
regulate the body, learn from a lifetime of
experiences, and summon the memories
and thoughts that are unique to each of us.

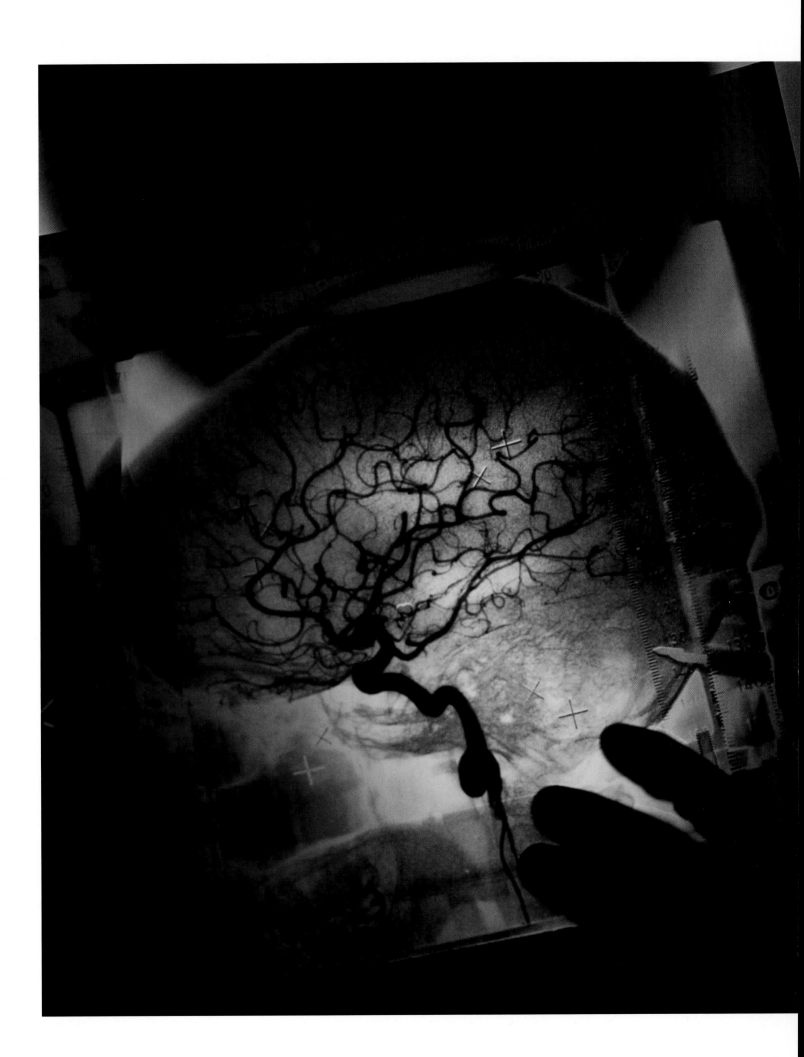

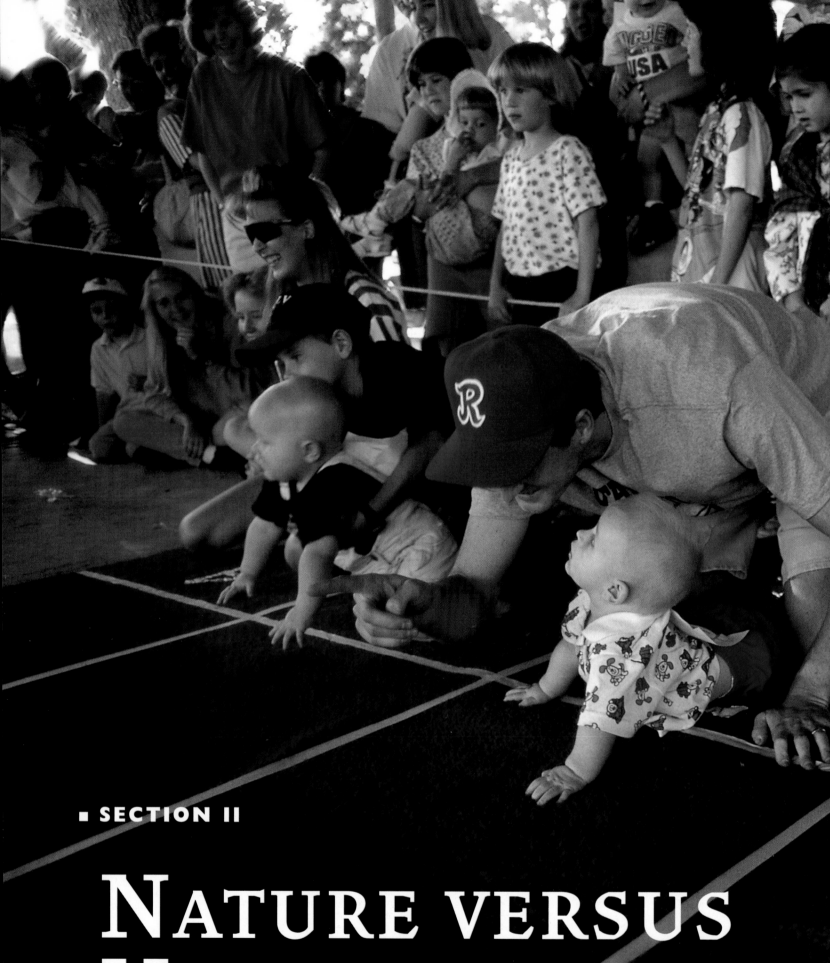

NATURE VERSUS NURTURE

GENES AND
THE BRAIN

No two people in the world are exactly alike. Along with other traits, people differ from one another in height and weight, skin pigmentation, eye and hair color—and that list doesn't even include the smorgasbord of personality and behavioral differences. Given such variations among people it seems natural to conclude that we all possess very different genes. But such a conclusion simply isn't true. Genetically, we are all very much alike.

Your genetic endowment is 99.9 percent identical to that of the stranger you passed in the street earlier today. Indeed it's only that remaining 0.1 percent—one part in a thousand—that distinguishes you from any person who has ever lived. Even more surprising, all genetic differences result from variations in only four basic components.

DNA, the hereditary material responsible for every structure and function in the human body, is composed of only four chemical bases, or nucleotides, called A, C, G, and T (for adenine, cytosine, guanine, and thymine. These four bases are the biochemical embodiment of our genes. Think of them as letters in a four-letter alphabet.

■ The genetics factor: In the Arkansas Ozarks, Ben Fruehauf plays checkers (or fuses) with Jack Elden Baker, a mentally retarded man with Down's syndrome, the result of an extra chromosome 21. PREVIOUS PAGES: Some contestants in this "baby race" will excel due to natural athletic ability; others will do well because they have learned to crawl quickly toward their mothers, who wait at the finish line.

Just as words are formed from the arrangement of letters, the A's, T's, G's, and C's that compose our genes are arranged into three-letter words. Those words are then arranged into sentences that serve as instructions for the construction of proteins within each cell of the body—including all of the brain's neurons and glia, along with all neurotransmitters and their receptors.

DNA's structural arrangement within the nucleus of each body cell is both simple and infinitely complex. The four nucleotides pair up in the form of an endlessly spiraling staircase. Your mother's genes comprise one of the banisters of that staircase; your father's genes comprise the other. At conception, one-half of the chromosomes from each parent arrange themselves into the spiral staircase pattern of the next generation.

So far, about two-thirds of our DNA's three-billion-unit code has been read by teams of scientists laboring on the Human Genome Project. Their goal is to arrive at the nucleotide sequence of each of the genes that make the proteins involved in both normal functioning and disease states. How many genes are we talking about here? Scientists can't be entirely certain, since only about 10 percent of DNA can be confidently identified as belonging to some gene. What is the remaining DNA doing?

The vast bulk of our DNA consists of strings of seemingly meaningless letters that fail to form words. And while such seemingly random patterns may yet turn out to transmit meaningful information, scientists haven't been able to translate them yet. As a result, no one is certain about the number of genes in the human genome. Recently the scientists working on the genome project have revised their estimates to about 50,000 genes.

But however many genes reside on our chromosomes, every cell in the human body (with a few exceptions, such as red blood cells) contains the genetic instructions for creating a complete human being. Most people are surprised to learn of this genetic egalitarianism among body cells. Certainly a neuron and a skin cell don't seem to have much in common. The neuron forms the basis for the human mind, while the skin cell performs the far humbler role of serving as a protective body barrier. The two cell types also differ enough in appearance that, after only a few minutes of instruction, even novices can easily distinguish one from the other. But appearances can be deceptive.

Neurons and skin cells only look and act differently because different genes are expressed in each cell. Moreover, only a small fraction of the total number of genes in each cell is ever expressed. In other words, turn on these genes and you end up with a skin cell; turn on those and you get a neuron.

Genes are turning out to be especially important in the operation of the brain. At least 30,000 of the genes known to exist within the human genome are expressed entirely or primarily in the brain. That's a significant portion of all the genes we have. Interestingly, about a third of all genetic defects affect the brain and nervous system. Scientists have arrived at these conclusions partly on the basis of what they have learned about the link between defective genes and defective brains.

Currently, neuroscientists can identify more than 200 genes that can cause or contribute to brain diseases. We now know, for instance, that a gene on chromosome 19 is associated with more than half

Specter of Huntington's disease haunts victim and scientist Nancy Wexler (right), whose mother died of the brain disorder. By charting genealogies in a region of Venezuela with a high rate of the disease, Wexler contributed to the discovery of the particular gene responsible.

of all Alzheimer's cases. An extra chromosome 21—three chromosomes instead of the normal two—results in Down's syndrome. Other diseases with genetic components include Parkinson's disease and several forms of epilepsy. One of the most destructive—and at the same time instructive—genetic brain disorders is Huntington's disease.

Neuroscientists have learned a great deal about genes and the brain from their investigations into Huntington's disease, the inherited brain disorder that killed folk singer Woody Guthrie, among many others. This illness—which afflicts about 30,000 people in the United States and puts another 150,000 at risk—provides a hint of how researchers will home in over the next decade on a host of genetic diseases.

Huntington's is particularly tragic because its first symptoms usually don't appear until its victims have reached middle age. Thus, in all likelihood, they may already have passed this genetic defect on to their children. For most of their lives they seem perfectly normal. But in their later years, over a period ranging from about 10 to 15 years, the victims of this cruel and merciless disease enter a downward spiral that ends in death.

First described in 1872 by George Huntington, a physician practicing in East Hampton, Long Island, the illness strikes one-half of the children of parents who have the affliction. Affected individuals of what once was called Huntington's chorea display disabling and involuntary movements that often bear some resemblance to a bizarre dance (chorea is Greek for "dance"). Accompanying these movements are slowly progressive neuropsychiatric symptoms

such as angry outbursts, an increased startle response, and impulsive or erratic behavior.

Eventually the Huntington's victim lapses into profound and irreversible dementia. The cause of this so far untreatable disease is a loss of neurons in the corpus striatum (a region of the basal ganglia, below the cortex, involved in the control of movement) along with other parts of the brain, including the cerebral cortex, the thalamus, the brain stem, and the spinal cord.

In the 1980s neuroscientists located a mutant gene for Huntington's on chromosome 4. After intensive efforts the nature of this mutation was identified. It consists of an unstable DNA segment containing an excessive number of three-letter "words" that repeat the sequence cytosine, adenine, guanine (CAG). The effect is similar to what happens when you play an old phonograph record that has a scratch: finally the needle gets over the scratched portion, the stuttering stops, and the song goes on to completion. Except with Huntington's, it's the music that stops. The numerous repetitions of CAG so interrupt the brain's normal "song" that they lead inexorably to the onset of Huntington's symptoms, the widespread death of nerve cells, and eventually the death of the victim.

Huntington's disease is only one of a growing number of brain diseases caused by a stuttering within different genes, due to repeated DNA triplet "words." And while Huntington's is rare, some other genetically based triplet-related illnesses are distressingly common.

For example, approximately 3 percent of the population is mentally retarded, possessing IQ levels of 70 or less. Several factors suggest a genetic basis for most instances of mental retardation. For one thing, more males than females are affected. A male preponderance would be expected in an illness resulting from a mutation on the X chromosome. That's because women have two X chromosomes and are therefore afforded some measure of protection from possessing a faulty gene on one of those chromosomes, since their second and presumably healthy X chromosome often serves as a corrective. Because males possess only a single X chromosome, however, they suffer the full consequences of any mutation involving that chromosome.

One way of testing whether or not an illness is genetic is to create a complete family history, with each family member traced back several generations. Such an analysis reveals that almost half of all X-linked mental retardation is caused by a single mutation, a gap observed on the X chromosome of affected individuals. Since that chromosome appears fragile when looked at under a microscope, geneticists refer to the associated mental retardation as fragile X syndrome.

The family tree of a person with fragile X syndrome typically reveals a mother who carried a mutant gene for the illness on one of her two X chromosomes, and passed that gene on to a son. The presence of that gene on his lone X chromosome caused him to come down with the illness. But even this fails to account for several intriguing aspects of fragile X syndrome.

As with Huntington's disease, fragile X involves a triplet repeat, in this instance cytosine, guanine, guanine (CGG). In the full-blown illness, that sequence repeats an exorbitant number of times, amounting to perhaps 800 or more CGG triplets. As a general rule for triplet-repeat diseases, more repeats mean greater severity—and earlier onset of the disorder. While a woman with 60 repeats has only a 15 percent chance of passing the illness on to a son, the risk climbs to 50 percent if she has 100 repeats. And the length of the repeated sequence gets longer with each generation. This increase forms the basis for what scientists refer to as *anticipation*—whereby a hereditary degenerative brain disease gets worse and/or occurs at an earlier age with each succeeding generation.

Not all individuals with the CGG (fragile X) abnormality are mentally retarded, however. In essence, the actual transfer and expression of the gene in the form of mental retardation varies from one family to the next. But over time, the increasing length of repeats through successive generations eventually results in half of the boys that were born to mothers who have the defective gene coming down with the disease.

Research has shown that development of fragile X syndrome becomes inevitable in male offspring when the triplet repeats exceed about 230. At that level, a specific gene is turned off, resulting in the loss of a particular protein that gene normally produces, called FMRP. As a result, the affected neuron can no longer manufacture many of the cell components needed to establish connections with other neurons. Without such connections, normal learning cannot occur. Fragile X researcher Stephen T. Warren, of the Howard Hughes Medical Institute, maintains, "The absence of FMRP is the cause of the learning problems and associated symptoms in children and adults with the fragile X syndrome."

The mechanism for Huntington's is slightly different. With this condition, the faulty gene leads not to the loss of one protein but to the formation of another, mutant protein. Each stutter in the gene

Lifelong epileptic, Russian author Fëdor Dostoevski kept logs of his seizures and experienced auras—visions—that he considered a creative gift. Epilepsy is a generic brain disorder with many different forms and symptoms. Like numerous other brain illnesses, it has genetic origins.

adds another molecule of the amino acid glutamine to the protein. As the number of glutamine links in the chain increases, the protein becomes stickier, causing it to clump and misfold—with normal proteins as well as with itself. The accumulation of the clumped material eventually causes the affected neurons to sicken and die.

"This process is quite slow, and will manifest itself clinically as a late-onset neurodegenerative disease when the neuron can no longer cope with the cumulative toxic effects of the mutant protein," writes Huda Y. Zoghbi, of Baylor College of Medicine in Houston.

Fortunately, such triplet-repeat diseases can now be accurately diagnosed through genetic testing. This lessens the despair and sense of inevitability associated with an incurable and devastating brain disease, and helps people make personal decisions concerning their future. "Genetic testing also allows researchers to study these diseases at their most fundamental level," says Zoghbi. "If we can understand these diseases at that level we will be better equipped to develop effective therapies."

Most neuropsychiatric illnesses involve several or even many defective genes, and picking out the bad ones isn't always easy. Sometimes, perfectly normal genes can be mistaken for faulty ones.

For instance, geneticists must be on the lookout for polymorphisms—a ten-dollar term that means, in plain English, that individuals with sharply distinct qualities may be perfectly normal members of a varied population. As an example, try to recall the last time you ate asparagus. Remember a peculiar smell, similar to that of boiling cabbage,

the next time you urinated? If you did, count yourself among the 43 percent of the population possessing the gene for this trait. The 57 percent without this gene fail to produce odiferous urine—or even to detect the odor if exposed to it. This is an example of genetic polymorphism. Those who don't produce or detect the smell aren't abnormal; they just happen to lack a specific gene. Other polymorphisms exist in regard to other odors and tastes. Some people perceive them, some people don't.

When discussing perceptual or behavioral differences among people, it's important to keep polymorphisms in mind. For instance, some people possess highly attuned musical talents; others are tone-deaf. Are the latter abnormal in some way? Probably not. The ability to appreciate music varies along a continuum in the population. So does sociability. Some people are social butterflies who are never really happy unless they're schmoozing. Others prefer their own company and socialize with the same enthusiasm they exhibit at the prospect of taking cough medicine. Are they abnormal? Not necessarily.

A genetic approach to brain disease must consider not only genetic polymorphism but also the underlying genetic complexity that characterizes illnesses affecting thought and behavior. Manic-

FOLLOWING PAGES: The brains of skilled musicians process music differently from non-musicians. But instruction and practice can help new learners—such as this pair of Russian boys, playing in time to their teacher's hand—modify their own neural circuitry and process music like real pros.

depressive illness (bipolar disorder), schizophrenia, autism, and the vast majority of other behavioral disorders result not from a single gene but from a large number of genes. Normally, genes work together like the instruments in a symphony. Just as a symphony requires the participation and coordination of a large number of different instruments, a healthy brain requires the timely expression of a host of different genes. If a gene fails to "turn on" at the normal time, the failure may disrupt the expression of other genes, just as a discordant note from a French horn can disrupt the beauty and majesty of a symphonic performance.

Steven E. Hyman, director of the National Institute of Mental Health, believes, "Mental illnesses appear to result from the interaction of a relatively large number of genes working in concert with as yet unidentified non-genetic factors." But Hyman also sensibly cautions about placing too much reliance on genetics. Genes, acting alone, are not the sole explanation for behavior. Strictly speaking, there is no "schizophrenia gene"—but a person carrying certain genes may be at increased risk of developing schizophrenia.

"Behavior, both its normal variants and mental illnesses alike, cannot be captured by the traditional terms of the 'nature versus nurture' debate," adds Hyman. "Personality, behavioral repertoires, and mental illnesses are neither the result of environmental factors alone nor the inexorable workings of isolated genes."

His point is supported by research on mental illness among twins. According to one Swedish study, the occurrence of bipolar disorder in both twins was higher with identical twins, which result from a single egg, than with fraternal, or two-egg,

twins. While this suggests genetic involvement, the rate is slightly lower than would be expected for strictly genetic transmission. This suggests that genetics accounts not for the disease itself, but rather for an inherited *vulnerability* to the disease. Genetic studies of schizophrenia have led researchers to a similar conclusion.

Inherited vulnerabilities occur for all sorts of conditions, not just mental illnesses. Each of us has our own Achilles heel (or heels), based on individual genetic and environmental variables. Consider a pair of identical twins. They look alike, often act alike, and share many other similarities. But to their parents or others who know them well, they differ in many ways. One child may tend to be more active and inquisitive than the other.

Geneticists believe that slight differences in environment or experience may influence some genes to "turn on" in only one member of any twin pair. Perhaps during early development only one member contracted an infection that required an antibiotic. That antibiotic (or perhaps the virus or bacteria responsible for the infection) could have altered the pattern of gene expression and thus produced a behavioral difference for that twin.

Certainly we know that environmental experience alters gene expression throughout our lives. Physical exercise, for instance, influences gene expression to enlarge heart and skeletal muscle cells. And if we drink too much alcohol, we alter the expression within our liver of important enzymes that metabolize alcohol. Other chemicals exert similar though subtler effects within the brain. If we are prescribed mood-altering drugs, they alter the transmission of

■ Every baby's genetic makeup determines its predilection for certain traits. But environmental experiences often shape the expression—or non-expression—of those genes. So it's not really nature versus nurture; it's both.

nerve signals and the expression of genes within specific neurons in our brain.

Even spoken words can result in changes in the expression of genes. At the simplest level, consider what happens when someone is subjected to daily criticism. The critics' words are transformed into nerve signals that travel along the auditory nerves through several way stations until finally reaching the auditory reception area in the temporal lobes. From there, signals rapidly disperse throughout the brain and activate specific neuronal circuits.

At the level of the synapse, the signals induce changes in neurotransmitter-receptor interactions on the nerve cell's membrane. These interactions, in turn, lead to the activation within the neuron of a series of messenger molecules, which travel to the cell nucleus, where they activate or repress certain genes. Imagine, for instance, that the criticized person happens to carry a gene predisposing him to depression, but, thanks to favorable circumstances so far in life, has never been depressed. Under the influence of mounting criticism, however, his "depression gene" is tripped off. In a person lacking

this genetic predisposition, it's unlikely that daily criticism alone would have led to a depressive illness.

We all carry genes that predispose us to certain physical and behavioral responses. Yet these responses may only emerge as a consequence of specific experiences. Traditional explanations for human personality and behavior emphasized either genetic inheritance (nature) or environmental experiences (nurture). But nature-nurture debates are no longer appropriate. Both genes *and* experience are important when it comes to understanding personality, behavior, mental illnesses, and other components of mind and brain.

Since genetics shares the stage with environmental experiences in the shaping of personality, predictive tests for neuropsychiatric illnesses aren't possible. No matter how "bad" a person's genetic inheritance, the likelihood of that person coming down with a brain disease similar to that of their parents or other relatives can only be evaluated statistically. While the odds of getting sick may be greater in one person compared to another, genetically speaking, that person may still live a perfectly normal and productive life. And isn't that a liberating, freedom-enhancing thought? Individual destiny remains primarily dependent on individual efforts, no matter what the odds predicted by genetics.

In short, human behavioral diversity is simply too great to be accounted for by genetic factors alone. We are not fruit flies or mice. Genetics is only one tool among many. Or, as NIMH director Steven Hyman cautions, "If we are to harness the power of genetics to help us understand human behavior and its disorders we are likely to find ourselves engaged in a complex, interdisciplinary effort that will use all of the tools of neuroscience."

Thanks to a quality called plasticity, the highly adaptive human brain can alter itself to partly make up for the loss of one sense by enhancing the powers of another. Thus many blind people, like this girl, develop a heightened sense of touch.

EXPERIENCE AND PERFORMANCE

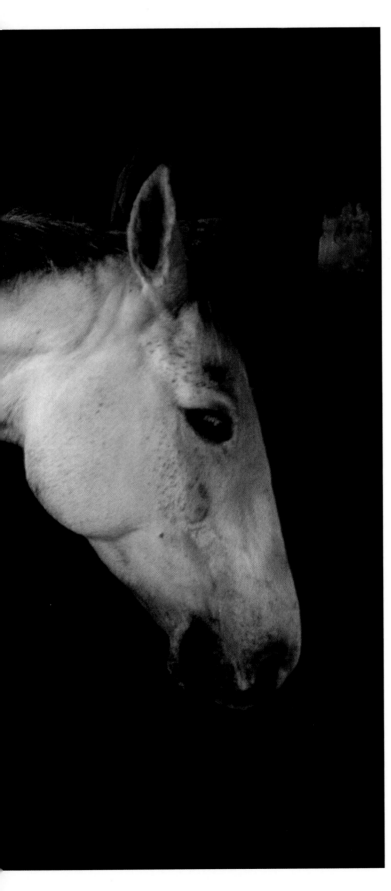

I magine that you're staring at a map of the United States. What you see is a static representation of our nation, with each state neatly distinguishable from its neighbors. Nothing much has changed over the past century in terms of state borders, and changes aren't likely in the foreseeable future.

Now compare that geographical map with the weather map you see on the evening news. Nothing is static there. Weather maps are dynamic, representing rapidly changing conditions.

For years the adult human brain was thought to be more like a geographical map than a weather map. Following an initial period of change and fluctuation very early in life, the mature brain was believed to settle in and assume a specific identity. Neuroscientists used to joke that if the appropriate technology ever became available, it would be possible to distinguish different people on the basis of the organization of their brains—a sort of neurological fingerprint.

Actually, a permanent "brain print" would never be possible, since the brain changes from moment to moment and represents a dynamic rather than a static structure (the weather map versus the geographical map). The first clue to this arrangement came from the observations of an American neurosurgeon named Wilder Penfield.

In the 1950s, Penfield carried out a series of experiments in the operating room at the Montreal Neurological Institute. With his patients awake on the operating table (many brain operations can be carried out on a conscious patient, since the brain itself doesn't contain pain receptors), Penfield

The homunculus: If our body parts were sized according to their representation in the brain's motor and sensory areas, this is what we'd look like. Huge lips, tongue, hands, and feet; tiny legs, arms, and trunk.

applied a gentle electrical current to different brain areas. With each stimulation, Penfield asked the patient what he experienced, and recorded the response. He quickly noted that the patients reported sensation in certain body parts more frequently than others. He then drew a map of the brain areas he had stimulated and correlated it with his patient's reports.

A key observation emerged from Penfield's operating room experiments. Functional importance rather than physical size determines the amount of brain tissue devoted to a particular body part. Thus those parts of the brain responsible for sensation and movement involving the mouth, lips, tongue, fingers, and eyes are much larger than those devoted to the small of the back, the trunk, or the stomach. From these observations, Penfield and others developed what would be called a homunculus (above), a drawing in which body parts are sized according to their representation in the brain rather than their actual, physical size. Because larger areas of the brain are concerned with the face, lips, fingers, and feet than the trunk, back, or legs, these body parts appear much larger.

The result is bizarre, almost like the image you see when standing before a fun-house mirror. Yet such an arrangement makes perfect sense. Speech and use of the hands are extraordinarily important to us, and involve far more complicated nerve and muscle interactions than any activities carried out by the trunk or back. Thus the brain is *functionally* organized—for us and for other species. The rat brain, for instance, devotes a large area of its sensory cortex to the representation of facial whiskers; in the raccoon, the paws are over-represented; in the platypus, the bill is the big winner.

This functional organization continues at lower levels of the nervous system as well. For instance, we have more touch receptors in our fingertips than other areas of our fingers. More motor fibers are present, per unit area, in the tongue than in the small of the back. Such observations lead to a chicken-or-egg dilemma. Did the basic organization of the human brain millions of years ago determine for us what we do best (working with our hands and speaking), or did the increased use of our hands, mouth, and tongue over hundreds of thousands of years result in the brain's current functional organization?

———————————

While no definitive solution is possible, Penfield's findings provided an important insight into the organization of the brain. Each part of the body is represented in the cortex at a specific location. Activation of a specific point on the brain with an electrode elicits sensations at the appropriate body area. Similarly, individual nerve cells in the brain increase their activity when the corresponding area of the body is touched.

Throughout our lives these body maps remain dynamic and potentially changeable. Neuroscientists have learned, for example, that if a monkey's arm is surgically amputated, the brain reorganizes itself. Those brain areas formerly responsible for sensation and movement involving the lost arm switch their allegiance and now mediate sensation and movement involving the face.

The same is true for humans. As a result, a single light touch to an amputee's face can result in phantom sensations in a forefinger of the missing hand. This also explains why persons who have lost

There is nothing rigid about the way the brain
represents its body or the objects around.
—Antonio R. Damasio

an arm or leg often describe a tickling or pain in the missing limb: The body maps within their cerebral cortex have become reorganized.

Nor is cerebral reorganization limited to traumatic situations such as amputation. Experiences early in life can produce dramatic changes in the circuitry of the brain and the arrangement of the homunculus. It's been discovered, for example, that among users of American Sign Language who have been deaf since childhood, the brain's visual centers take over many of those functions normally performed by the auditory areas. Researcher Martha Constantine-Paton of Yale University showed me a functional MRI picture of a deaf person's brain response to visual motion. I observed activity spreading from the visual cortex forward to encompass large portions of the cortex including the frontal lobes.

"Activity patterns within both auditory and visual regions of cortex are changed dramatically following this altered early auditory experience and these changes are not seen in individuals who learn sign language as an adult," said Constantine-Paton.

Among the deaf who have used sign language since childhood, the brain's language-recognition centers respond not to spoken but to signed words. This response is so firmly established that artificial stimulation of the auditory pathways that connect the ears to the brain fails to produce any effect on the language recognition centers. Only signed words activate these regions.

———————

Similar brain dynamism can be observed among blind people who have learned to read by means of the tactile alphabet invented by Louis Braille.

Although each finger has a distinct group of cells representing it in the brain, the map of brain cells for each finger differs among blind readers of Braille. The reading finger—and only that finger—becomes represented by many more brain cells. In contrast, the representation within the brain of the other four fingers doesn't differ from that of sighted people. Of course, the enhancement of the brain's representation of the reading finger carries practical consequences. When a blind person reads Braille, the subtle impressions made on his fingertip by the tiny bumps on the page take on special meaning, thanks to the brain's increased representation for that finger.

But if a blind person stops reading Braille even for a short period of time, his brain map undergoes a dynamic re-transformation. Usage is extremely important; a blind person's brain map is larger after six hours of reading Braille than it is after a day off. From such observations, neuroscientists have concluded that the brain recruits more cells when a particular job needs to be done but reassigns those cells when the job is over.

———————

Yet even when they're not reading Braille, many blind people possess a finely tuned tactile sense, which enables them to make subtle distinctions by touch alone.

A similar enhancement in the tactile sense commonly occurs with fashion designers, for example, helping them differentiate textures and contours. The designer and the blind person make these distinctions based on the presence of an enlarged brain area dedicated to the touch receptors at the fingertips.

Based on these and other findings about the plasticity of neuronal connections among the deaf and the blind, new treatments are now becoming available for disorders previously considered untreatable. For instance, children with specific language impairment (SLI), a speech disorder responsible for many cases of dyslexia, have so much difficulty distinguishing consonant-vowel syllables such as "ba" and "da" that they quickly lose the thread of a conversation.

Treatment of SLI consists of a computer program that alters the acoustics within speech syllables. The computer artificially extends "ba" and "da" until the child can successfully discriminate between them. At this point, the child instructs the computer to speed up the syllables until his or her consonant-vowel discrimination ability approaches that of a normal child. Although no one can be certain what's happening, the neuroscientists who developed this method believe it alters the sensory map in the cortex.

––––––––––

Other examples of brain plasticity abound. Take musicians: The motor cortex—the region that controls hand and finger motions—is much larger in professional keyboard and string musicians than in non-musicians. Not only is the motor cortex of each hemisphere larger, but so are the white-matter fiber tracts that connect the cortices of both hemispheres together. Most intriguing of all, these motor cortical enlargements are related to the age at which the musician started training. The earlier in life one starts playing, the bigger the motor cortex. Even so, it's never too late to enhance your brain by taking up a musical instrument. Whatever your age when you begin playing, the brain areas given over to music will become larger than observed in people with no musical inclinations.

Some brain alterations take place instantaneously. "Moving one finger rapidly and repetitively for a few minutes alters the brain representation of that finger for a while," writes Antonio R. Damasio, a neurologist and researcher at the University of Iowa College of Medicine. "There is nothing rigid about the way the brain represents its body or the objects around."

Again, the brain is not a static geographical map but more like a fluid and dynamic weather map. Moreover, this dynamism extends over our entire life span. Whether we are one day old or 100 years old, our brains are undergoing dynamic changes all the time. Each daily activity shapes the brain, sculpting it into a different form depending on the nature of that activity. The brain changes from moment to moment, whether we like it or not.

Still, we do have quite a bit to say about just how our brain will change. As we'll see later in the discussion on aging, we can passively allow our brains to atrophy, due to a lack of interest in and connection with the people and events in the world, or we can enhance the wonderful complexity of the brain we have. The choice is ours.

■ For those with or without sight, reading and writing in Braille enhances the sensitivity of the fingertips and reorganizes the touch receptors in the brain. This process is so dynamic that some loss in sensitivity occurs after even short periods away from communicating in the Braille alphabet.

VIVE LA DIFFÉRENCE!

"It's a woman's intuition." "Men are so stubborn." "It's a man's world." Not so long ago such expressions were part of everyday discourse. Now most of us are justifiably uncomfortable with stereotypes that imply men and women can be neatly set apart solely on the basis of emotional or behavioral traits, personalities, or propensities. We recognize that gender differences, if they exist, must—like all forms of inheritance— depend at least partly on environmental influences.

Yet those who have spent time with children aren't entirely comfortable with the claim that innate, biologically based gender differences don't exist. Many recognize—or at least think they do— that from infancy onward, males and females often behave and think differently when faced with similar situations. But if such observations are valid, a nagging and socially explosive question persists.

Are there recognizable and meaningful differences extending over a lifetime between the brains of boys and girls, men and women? Two decades ago, to even ask such a question was to risk being branded a sexist. In my first brain book, *The Brain: The Last Frontier*, published in 1979, I mentioned some emerging work on brain-sex differences in humans and concluded, "On the basis of the information available it seems unrealistic to deny any longer the existence of male and female brain differences." That seemingly innocuous remark elicited quite a response! I became engaged in several spirited discussions with readers who thought I should have omitted any reference to the subject.

Sexual differences in the brain are controversial partly because sexual identity and behavior in our

■ Teenage ballerinas-to-be practice their moves at a French conservatory. Although environmental influences play a major role in gender differences, so does genetic makeup; the brains of boys and girls are *not* the same.

species aren't nearly as neatly compartmentalized as in the creatures (principally rats and mice) that served as early research subjects for probing the existence of brain-sex differences. Indeed, when applied to humans, the very word *sex* has several different meanings.

Consider Jan Morris, a widely recognized and gifted travel writer. At one time Jan was a man, James Morris. After transsexual surgery and hormone treatments the former Mr. Morris became Ms. Morris.

At the genetic level—what we call genotypic sex—Jan Morris remains a man, because he still has an X and a Y chromosome in every body cell. Nothing can change that biological identifier. But in terms of physical characteristics, or phenotypic sex, James Morris no longer exists: The operation removed James' male sexual organs, while hormones and cosmetic surgery have feminized the body contour. Were you to meet and talk with Jan Morris, it's likely that you would never suspect that Jan was once James. In terms of gender identification, Jan no longer thinks of herself as a man but as a woman. So is Jan really a man because of his immutable genetic composition, or is she a woman because of her altered anatomy and gender identification?

Obviously, genotypic sex, phenotypic sex, and gender identification aren't always aligned. A man who feels that he is actually a woman may marry, father children, and even live out his whole life within a false gender role. Or the mismatch between appearance and psychological identity may become so agonizing that, as with James Morris, the person may undergo surgery to bring into alignment the perceived sexual mismatch. Sometimes, false gender roles are imposed on children.

In his book, *As Nature Made Him: The Boy Who was Raised as a Girl*, author John Colapinto describes the travails of a male infant, Bruce Reimer, who was accidentally penectomized during a routine circumcision. Due to this tragic medical error, Bruce's parents were advised to raise their child as a female. Bruce underwent castration and the construction of a cosmetic vagina. In place of boy's toys, Bruce—now Brenda—was supplied with dolls and dresses, and was tutored on how to talk, walk, and generally act like a girl. But the anticipated transformation never took place. After learning at age 14 what had happened, Bruce decided to reclaim his male sexual identity. Today he is married and the adopted father of his wife's children from her first marriage.

What, then, determines one's sexual identity? Jan Morris and Bruce Reimer possess the same genetic makeup, but each has chosen a different sexual identity. Obviously, the presence of an X and a Y chromosome in each cell of an embryo does not provide the final word. Nor do two X chromosomes necessarily guarantee female gender identity. Some genotypic women elect to undergo operations aimed at turning them into phenotypic men (although this occurs more rarely than the man-to-woman switch).

Despite such variations, sexual development in most people generally follows a clear-cut pattern. The genes (genotype) determine the physical characteristics of the sex organs (phenotype), which in turn produce most of the sex hormones. During even the early stages of development, the testes begin making androgens and the ovaries begin making estrogens. This difference in circulating hormones leads to different developmental patterns

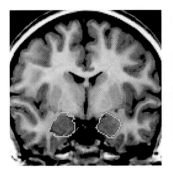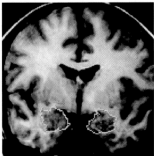

MRI scans of an 8-year-old girl (far left) and boy (left) reveal differences in the structural development of the amygdala and other brain areas. The amygdala is bigger in boys, relative to total gray matter. But by adulthood, women possess a relatively larger amygdala.

throughout the body, including the brain. These same hormones also influence behavior.

In animals, a wide range of behaviors are sexually defined. They run the gamut from distinct male-female postures and responses associated with sexual intercourse, to distinct parenting roles. Part of this sexual dimorphism can be traced directly to the influence exerted by the Y chromosome.

The Y chromosome of male mammals carries a special gene that produces a protein that promotes the formation of the testes, the production of testosterone, and the subsequent development of the male brain. In the absence of this gene, a default developmental program springs into action, leading to female brain patterns, sexual characteristics, and some aspects of behavior.

Scientists have speculated for years about genetic influences on brain structure and behavior. It seemed reasonable to suppose that certain behaviors in males and females might result from male-female brain differences. But what are those differences and how does one demonstrate them?

In the late 1970s Roger Gorski and his colleagues at the University of California in Los Angeles first described easily discernable differences in a specific nucleus in the rat hypothalamus. Called the sexually dimorphic nucleus, or SDN, it is small in females and large in males. Moreover, it can be enlarged in females by early injections of androgens, and its size can be decreased in males by castration.

Since Gorski's original observations, neuroscientists have turned up other sexually distinguishing hypothalamic nuclei. Studies of rhesus monkeys have revealed that some neurons fire rapidly during

sexual approach but decrease their firing rate upon contact and mating with the female partner. Other neurons only millimeters away remain quiescent until mating actually occurs, then continue firing vigorously during intercourse. Thus, in monkeys, some hypothalamic neurons are dedicated to arousal, while others fire only during copulation. Can such research be applied to human sexuality?

Many of the conclusions on sex differences in the human brain are based on analogies to rodents, rhesus monkeys, and other non-human primates. Not surprisingly, many people question the validity of these analogies, and considerable controversy exists in regard to the nature and extent of sexual differences in the human brain. There is, however, some agreement concerning observed differences in a series of four nuclei known as the INAH, or interstitial nuclei of the anterior hypothalamus. According to researcher Simon LeVay, one of these nuclei (INAH-3) appears to be larger in heterosexual men than in women—or in homosexual men.

Although LeVay's findings are provocative, most neuroscientists remain cautious about concluding that hypothalamic nuclei are responsible for sexual orientation and gender identity. The same holds for the recent discovery, by researchers at Johns Hopkins University, that the inferior parietal lobule of the brain is 6 percent larger in men than in women. This is the brain region associated with interpreting spatial relationships, examining mathematical problems, and estimating time and speed. But this doesn't mean that men are automatically better at processing these functions than women are. According to one of the project's researchers, Godfrey Pearlson, the gender-associated difference in size of the inferior parietal lobule is only apparent when assessing large populations and may not apply to specific individuals.

Dale Purves and other members of the Department of Neurobiology at Duke University Medical Center summarized the difficulty in interpreting the evidence for brain-sex differences in their textbook *Neurobiology*.

"Taken together, this evidence suggests that one way to explain the continuum of human sexuality is that small differences in brain structure generate significant differences in sexual behavior…. [But] as attractive as this hypothesis appears, the development of sexuality in humans is probably a good deal more convoluted."

One of the reasons for caution is the difficulty involved in replicating earlier findings. Even more important, anatomical differences may reflect only correlations, rather than cause-and-effect explanations. Just because one or another nucleus in the hypothalamus may differ among men and women, or homosexuals and straights, does not necessarily dictate sexual behavior or orientation. As a result, many neuroscientists are turning their attention from brain structure to behavioral differences between the sexes. The findings are intriguing.

Among children, for example, more boys than girls are afflicted with severe mental retardation (1.3 boys for every girl), speech and language disorders (2.6 to 1), learning difficulties (2.2 to 1), dyslexia (4.3 to 1), and autism (4 to 1). What could account for these gender differences? One theory holds that male fetuses are more likely than female fetuses to invoke a foreign-body response by the mother's immune system; the resulting hostile environment leads to fetal brain damage and a resulting increase in brain-behavioral disorders.

Testosterone's effect on the developing brain may explain why men are drawn to aggressive, physically punishing sports. But physiology isn't absolute: Female soccer, lacrosse, basketball, and other teams are now commonplace. Will football be next?

For spatial tasks you need more white matter than is available to most women.

—Ruben Gur

Another theory postulates that an unknown factor present during the last trimester of pregnancy slows cortical maturation in males, most significantly in the left hemisphere. Since girls are not affected by this mysterious factor, their normally maturing brains are better able to withstand the stresses of pregnancy and birth. This may explain why female fetuses experience quicker and more complete recovery than males after fetal brain damage. Nor is this female advantage limited to early stages of development. Following strokes within the left hemisphere, women suffer language difficulties less often than do men with the same brain damage.

———————

Because any discussion of gender differences in brain organization and function is bound to cause controversy, it helps to keep in mind that such differences do not imply that one sex is superior or inferior to the other. *Vive la différence!* Remember, many observed differences between the sexes are based on averages. Just as it's true that, on average, men are taller and heavier than women, it's also true that exceptions abound. Many of us don't conform to supposed brain-sex differences any more than we conform to stereotypes about height and weight.

On average, however, women tend to perform better than men on verbal tasks. They're also better at certain motor programming skills and at rapidly scanning the environment for selected features. Men, in comparison, tend to excel on some spatial abilities, especially those involving mental rotation. They also do better on mathematical reasoning and on motor skills directed at distant targets.

But even granting that these differences may hold true on average, how do we know they are not all cultural? Again, much of the evidence for inherent, hormonally based sex differences comes from rodent research. Male rats tend to work their way through mazes better than females. When males are castrated, or females are given male hormones shortly after birth, the performance patterns reverse. In such experiments, rearing practices and different life experiences are likely to be less important factors than early hormonal influences.

"It is not unreasonable to conclude that in humans also, much of the variation in spatial ability that we see across individuals, regardless of sex, may be related to such hormonal influence," believes Doreen Kimura, an expert on brain-sex differences now working in the department of psychology at Simon Fraser University, in British Columbia. Dr. Kimura readily admits "the precise neural mechanisms responsible for these differences in cognitive functions are at present a matter for conjecture."

Ruben Gur, of the University of Pennsylvania, has found that male and female brains differ in their ratio of white to gray matter; women generally have a higher percentage of gray, while men have more white. Gur also sees a correlation between the volume of white matter and a person's ability to perform spatial tasks. Because his top performers were all men who had higher volumes of white matter than did any of the females in his sample, Gur concluded, "for spatial tasks you need more white matter than is available to most women." How does one explain women who consistently outperform men on tests of spatial ability?

It's difficult to say for certain, but we do know that anatomy of the brain is only half the story. Hormonal influences also play a major role in the establishment of brain-sex differences. If a woman is treated with estrogen after menopause, her brain will behave like the brain of a younger woman when it comes to reading and memory. Estrogen increases activation in brain regions responsible for storing phonemes, the critical spoken sounds used in all languages to convey meaning. The person who did this research, Sally Shaywitz of Yale University, believes "estrogen may be related to reading and thus may explain why women are generally more verbal than men."

Estrogen may also play an important role in memory. Women deprived of estrogen experience memory lapses that disappear when the hormone is replaced. Barbara Sherwin of McGill University, in Montreal, studied 100 women in their mid-forties after surgical removal of their ovaries and uterus. Half of the women were treated with hormone replacement therapy (HRT); the other half got a placebo. The HRT group showed no change in memory function while the placebo group complained of memory problems and for the first time in their lives started writing out lists of things to be done. On a test involving reading and remembering words, the HRT women did as well after their operation as they had before, while the women on placebo performed significantly worse—a nice confirmation by Sherwin of Shaywitz's findings.

We're left with a major challenge. How can we reconcile the brain-sex differences discovered by neuroscientists with women's recent social and political advances? Today, there are more women engineers, physicists, architects, and mathematicians than at any time in history. It seems reasonable to assume that the surge of females in these formerly male-dominated professions has resulted more from changes in societal attitudes rather than any alterations in the organization of the female brain. Yet the findings of brain-sex differences are impressive and not easily dismissed. Let me suggest a resolution to this dilemma.

Assume for the sake of discussion that male and female brains are predisposed by anatomy and even chemistry to certain cognitive advantages and disadvantages. In a closed system such as a machine, these differences would determine performance for the life of the machine. But, as we learned in the previous chapter, the brain isn't a closed, unchangeable system. Rather, it's a highly malleable one that can be dramatically influenced by experience.

"Experience," by the way, includes not only what is encountered in the outside world but also such things as mental attitude and determination to achieve one's goals. All aspects of experience can alter brain circuits and induce changes that may diminish or even reverse the cognitive organization originally established by biology. Do I have any proof for such a metamorphosis? No. But such a proposal is based on sound principles of brain organization and function.

If the brain changes from one moment to the next—as we now know it does—why shouldn't it be able to form new circuits that bypass or modify established ones and thus overcome biological predispositions? If a person can reorganize something as seemingly immutable as sexual identity (as Jan Morris has), surely it should be no surprise that thousands of women and men have achieved occupational and professional success by modifying biological influences on their brain's cognitive and behavioral organization.

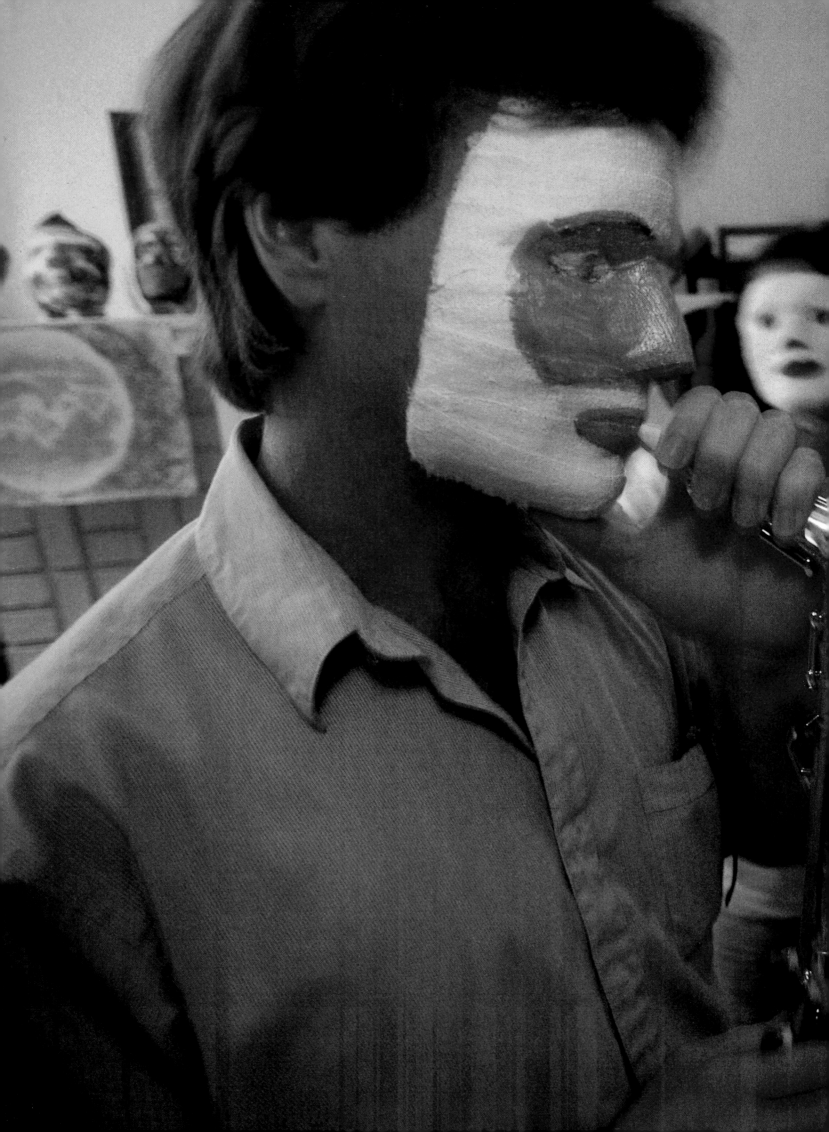

■ SECTION III

STATES
OF
MIND

Consciousness and Cognition

Let your mind just drift for a few seconds; don't try to think of anything in particular. Keep at it as long as you can. If you're like most people, thinking "of nothing" will turn out to be more difficult than anticipated. Within a few moments you'll experience certain thoughts, maybe even an urge to stop "wasting time." Eventually, despite your best efforts, your attention will gravitate to something happening around you or to some aspect of your mental activity.

Now imagine that, in the next room, some neuroscientists are using a highly sophisticated imaging device to monitor the moment-to-moment operations of your brain. As you allow your mind to drift, they observe a decrease in brain activity, particularly in your frontal lobes. But as your attention is captured by some object or thought, your frontal and prefrontal areas "light up," as do areas involved with your arms and legs—you're becoming restless, squirming in your chair, in anticipation of getting up and getting on with your day.

There are important differences between your experience of your own internal state and the experience of the people observing the operation of your brain. They notice changes on their futuristic brain monitors that correspond to shifts in your brain's activity. Based on certain brain potentials, they are even able to predict within milliseconds when you are about to take some action. In a sense, they know about it before you do. Yet they lack access to exactly what that action will be; they haven't a clue about your feelings of boredom and restlessness that eventually cause you to move.

Note that these two vantage points cannot be merged. No amount of conscious introspection on your part will teach you a thing about your brain as a physical object, even though consciousness is a property of your physical brain. Should you fall asleep or receive an anesthetic, your resultant loss of consciousness can be monitored by others through chemical and electrical alterations in your brain. Yet you are not aware of any of these chemical or electrical changes.

On the other hand, those scientists in the next room have no access to your consciousness, no idea what you may be thinking. While the mind and brain form a unity, our knowledge about them rests upon an irreducible duality. As philosopher Colin McGinn put it in *The Mysterious Flame: Conscious Minds in a Material World*, "You can introspect till you burst and you will not discover neurons and synapses and all the rest; and you can stare at someone's brain from dawn till dusk and you will not perceive the consciousness that is so apparent to the person whose brain you are rudely eyeballing. Even high-tech instruments only give the physical basis for consciousness, not consciousness as it exists for the person whose consciousness it is."

■ André Rouillard's surreal *Essor* (Flight) hints at the mind-brain dualism suggested by René Descartes. Consciousness—a state so obvious to the conscious yet so difficult to define—remains a fascinating mystery to the anatomist and physiologist. PREVIOUS PAGES: By adulthood, most of us become skilled at hiding our feelings behind social masks. A group therapy session in Orlando turns the tables, as participants don masks they have painted to express their true selves.

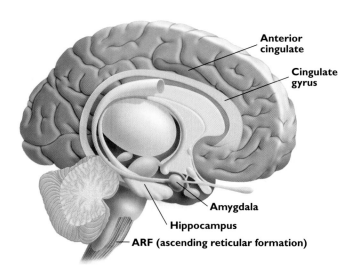

Anterior cingulate

Cingulate gyrus

Amygdala

Hippocampus

ARF (ascending reticular formation)

So far, neither neuroscience nor philosophy has come up with a solution to this mind-brain duality. Earlier we mentioned a few of the theories espoused over the centuries. Today we are no closer to an understanding of the mind-brain conundrum than our forebears, and can only agree with Thomas Huxley's 1886 observation, "How it is that anything so remarkable as a state of consciousness comes about as a result of irritating nervous tissue, is just as unaccountable as the appearance of the djiin, when Alladin rubbed his lamp…."

So, rather than try to present answers that no one has, let me suggest some approaches, and allow you to make up your own mind-brain about this dilemma. Suppose that, while in a discussion with a friend, you say something that isn't correct. Immediately after speaking, you have a feeling that your statement wasn't quite right, and you correct the error just as your friend tactfully points it out to you. How did you become aware of your mistake?

Error detection is something that we all do every day. Frequently after we've made a mistake, a kind of warning alarm bell goes off within our heads (not always immediately, of course, and sometimes never). Neuroscientists have located this error-detecting alarm bell in an area toward the front of the brain called the anterior cingulate.

Under ordinary conditions, the anterior cingulate is activated whenever we have to focus and concentrate our attention. Suppose you are asked to quickly come up with uses for a list of words presented one at a time. Each time you make a correct response, your anterior cingulate will "light up," showing a particular activation pattern.

Then, in a second part of the experiment, the same words are repeated in a different sequence. You find that both the mental effort required to perform the associations and your reaction times drop dramatically. In addition, the anterior cingulate is no longer activated.

Then a new word list is presented and you are challenged to come up with appropriate associations. Again your anterior cingulate springs into action. In short, its activity is greatest during times that require thought; as things become routine, activation of the anterior cingulate is reduced.

A similar activation occurred when you first became aware of your error in that conversation with a friend. That internal "alarm bell" corresponded to activation of your anterior cingulate. Thus, the same brain structure that's involved with active problem-solving—thinking—is involved with error detection. This suggests, among other things, that the anterior cingulate is a supervisory system important to consciousness. Examples drawn from neurology support this contention.

Patients with damage to the anterior cingulate do not initiate any action or even speak on their own. They sit for long periods of time without speech or movement, in a state called akinetic mutism. One patient who recovered from such an injury explained her previous passivity by saying, "Well, nothing ever came to mind." Only after her recovery did she become fully conscious once again and experience herself as the initiator of her own thoughts and behavior.

Even so, the anterior cingulate is not the "seat of consciousness" sought over centuries by philosophers and, more recently, by neuroscientists. There is no single brain area responsible for consciousness;

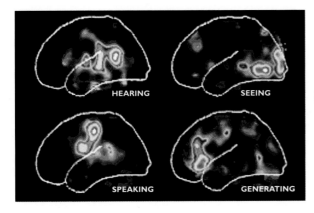

Four different activities with words— hearing them, seeing them, speaking them, and generating them—activate different areas in these PET scans, indicating that the brain is functionally organized, with different areas devoted to various tasks.

consciousness is not an entity but an active process that requires the participation of many components.

As a first condition for consciousness one must be awake and alert. This is the task of a complicated network of neuronal circuits located in the core of the brain stem and extending upward from the medulla—just above the top of the spinal cord— to the midbrain. Known as the ascending reticular formation or ARF, this network of neurons sends widely branching axons upward to distant parts of the brain. Within the ARF are clusters of neurons that manufacture and transport the important neurotransmitters norepinephrine, serotonin, and acetylcholine. The most basic function of this tightly compact yet specialized region is to keep us alert and awake. Permanent damage here can result in irreversible coma.

But consciousness requires more than mere alertness. We have to be able to smell the roses—to employ our senses in monitoring the world around us. And we have to be able to scurry across a crowded street or respond in countless other ways when our world suddenly changes. None of this is possible unless we can direct our attention—through the anterior cingulate and other areas—and access our memories for preceding events. At the same time, we must be able to form some kind of mental picture of the likely consequences of our actions. Finally, we must be able to "get in touch" with ourselves via internal monitoring—that sense of self and identity that separates each of us from everyone else in the world.

When Descartes declared, "I think, therefore I am," he gave short shrift to some very vital processes:
• Without his ARF, Descartes would not have been awake and therefore could not have reached his historic conclusion. The neurons in his brain stem bathed the neurons in his cerebral cortex with activating neurotransmitters. Without those neurotransmitters Descartes could not have integrated his internal experiences into the "I."
• And without the hippocampus on each side of his brain, Descartes would not have been able to remember and thereby maintain the continuity over time needed to formulate the concept of "I."
• Without normally functioning frontal and prefrontal areas, Descartes could not have gained the necessary perspective to experience himself as separate from the continual barrage of internal and external events that made up his world.
• And if his anterior cingulate weren't working, he wouldn't have been able to focus and concentrate his attention.

———————————

Yet this analysis of the neurology of Descartes' brain isn't completely satisfying either. Somehow it seems overly simplistic to imply that one of the world's great thinkers can be explained by simply referring to neurons and networks. Isn't there something else? But what is that something? And how should we think about it and describe it?

Philosophers have wrestled unsuccessfully for centuries with the so-called mind-body problem. Neuroscientists now claim some success, over a much shorter span of time. But, to be fair, they have loaded the deck by defining consciousness strictly in terms of the brain.

Philosopher John Searle, of the University of California, Berkeley, considers consciousness to be "a biological feature of the human and certain animal brains. It is caused by neurobiological processes and is as much a part of the natural biological order as any other biological feature."

As a neurologist, I'm comfortable with Searle's view. No doubt that's because over the years I've personally observed the effects of various kinds of brain damage on consciousness. The changes can vary from a slight decrease in alertness and wakefulness to deep and irreversible coma. In between these extremes, I've encountered people who deny their paralysis and seem unaware that they have suffered brain damage; others who have developed profound amnesia for important events in their past; still others who can't recognize their spouses or children by sight alone but only after hearing their voices. In such instances brain, mind, and consciousness seem inseparably interwoven.

There's another aspect of Searle's comment that I find especially intriguing: Is consciousness a biological feature of *certain animal brains?* Are some animals conscious in the same way we are?

Such questions aren't easily answered. For starters, we have to admit that it's difficult getting a handle on consciousness in other people. Someone once described a person seeking to define consciousness as a blind man in a dark room looking for a black cat that isn't there. Despite the fact that everyone reading this book knows first-hand the experience of consciousness, when they attempt to define it they become like that blind man stumbling in the darkened room. Each of us experiences consciousness but must take on faith the claims by others that they, too, are conscious. Whenever we look into ourselves, we encounter what the Austrian philosopher Ludwig Wittgenstein called the "beetle in a box" that we alone can see. No one can enter the consciousness of another.

When seeking consciousness in another species, the going gets even more complicated. While I can ask you about your internal experiences and you can describe them in words that make sense to me in terms of my own internal experiences, such an approach simply isn't possible with animals. Even so, students of animal behavior have devised some ingenious methods for indirectly assessing the possibility of animal consciousness.

One method, which involves nothing more elaborate than a mirror, was initially used to study children; an experimenter would apply rouge to a child's face and let the child look at itself in a mirror. The response varied according to the child's age. No child younger than a year would touch its face. Between 15 and 18 months of age, only one out of four children tested would reach towards the colored spot. But three out of four two-year-olds reached up and touched the rouged area.

Applying this experiment to other primates, only the chimpanzee seems to recognize itself in a mirror. Chimps respond at about the level of an 18-month-old child, reaching toward the mirrored reflection of the rouge. Some ethologists believe this rudimentary form of self-awareness indicates consciousness. But I wonder if such a far-reaching conclusion is really justified. Self-recognition, even in a young child, does not imply the existence of an internal dialogue along the lines of, "That's me there in that mirror; that's what I look like; that's the reflection of the *me* I've always experienced."

But before we read too much into mirror experiments, we might heed the advice of 19th-century biologist Lloyd Morgan. He suggested that one should always look for a simple, mechanistic explanation for even seemingly complex behaviors.

Consciousness is a biological feature of the human and certain animal brains.
—John Searle

(If you don't, you can wind up believing that automated doors and supermarket scanners exhibit some degree of consciousness.) In addition, mirrors are not part of the everyday world of jungle animals. More revealing are the responses of animals to the everyday events that happen in their realm.

Baboons will recognize and respond to the cries of other baboons, according to hierarchy and family relationships. Individuals seem to recognize their place in the community. Thus they display a primitive but serviceable sense of self that tells more about the mind of this animal than any artificially contrived experiment with mirrors.

Even more intriguing is a series of experiments involving deception and anticipation of the intentions of others. In one test, food is hidden while one chimp looks on (the witness). Another chimp (the bystander) does not know the food's location. As the witness moves around the cage, the bystander tends to follow, as if aware that the witness has some special knowledge. On occasion the witness even appears to deceive the bystander by leading it to areas that have no food.

In another example of deception, a juvenile baboon in the wild was observed to approach a female adult feasting on tubers. The youngster sounded a distress call even though it was in no danger. When its mother appeared, she drove away the falsely accused female adult, and the juvenile rushed to the prized meal. Does this prove that the youngster intended to deceive? Or do we read too much into a situation? Perhaps the juvenile had been genuinely frightened, and noticed the meal only after the stranger had left.

Other explanations for animal deceptions are also possible; chimps that live outside their natural habitat and interact constantly with people can be more likely to become sensitive to the thoughts and intentions of others. If so, then their apparent self-awareness may be no different than what occurs in human infants; their earliest rudimentary sense of self-awareness evolves as a result of the attention bestowed on them by parents and relatives.

Today the issue of animal consciousness remains undetermined. While some form of rudimentary self-awareness is possible for non-human animals, they are unlikely to possess the complex consciousness we have. The world has yet to witness a chimpanzee communicate the chimp version of "To be or not to be."

Much of our uncertainty about consciousness can be traced to difficulties in understanding the unconscious. And I do not mean what Sigmund Freud popularized—an unconscious filled with the bugaboos of sex and violence. Modern cognitive psychologists use the term far less judgmentally, simply referring to mental activities that never enter our consciousness. Oxford philosopher Stuart Hampshire sums it up:"A great deal of our thinking proceeds without conscious awareness. In the exercise of the use of language itself and in many of our skills we are thinking preconsciously, working things out without knowing how we worked them out, or by what steps we arrived at the conclusion."

Another philosopher, Bertrand Russell, offered a specific example:"Suppose you are out walking on a wet day and you see a puddle and avoid it. You are not likely to say to yourself:'There is a puddle; it would be advisable not to step in it.' But if somebody said,'Why did you step aside?' you would answer,

One of a host of brain-altering influences, music can powerfully affect one's state of mind—as seems obvious from the expressions on display at this Memphis blues concert.

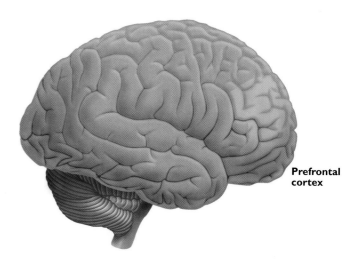

Prefrontal cortex

'Because I didn't wish to step into that puddle.' You know, retrospectively, that you had a visual perception, and you expressed the knowledge in words. But what would you have known, and in what sense, if your attention had not been called to the matter by the questioner?"

The vast majority of our mental operations are similar to Russell's puddle: We respond to most of the events and people around us without thinking consciously about them. Imagine the torture of driving a car if each and every action had to be consciously attended to. Instead, we learn to 'automate' frequently practiced routines so that our conscious awareness can work on other, more interesting things. Indeed, streams of perceptual, semantic, and behavioral processes regularly occur without conscious awareness. Yet these hidden mental processes regularly exert an influence on our behavior.

For instance, during a type of experiment known as masked priming, words are presented on a screen so briefly that they escape conscious detection. Nevertheless, this presentation increases the likelihood that related words will be detected. For instance, the word "mother" will be recognized much more quickly if it is preceded by the word "child." Brain-imaging studies show that masked stimuli—undetected by our conscious mind—exert a measurable influence on electrical activity and blood flow within the brain. Even though the subject is consciously unaware of the effect of the previous masked word, his brain responds differently from that of a person not exposed to the hidden word.

In another example involving PET scans, experimenters flash a series of nouns on a screen while the subjects are asked to rapidly generate appropriate verbs. Participants are tested both before and after practice, and the PET scans show marked differences in neural circuitry activation. In essence, more conscious effort is required to deal with novel stimuli, and this effortful response can be measured by functional brain-imaging techniques.

———————

Having come this far, you can probably guess the area of the brain that is most active in effortful consciousness: the prefrontal cortex. Whenever we attempt to hold something in our mind's eye, working memory is called into play. Working memory enables us to maintain an active representation of something that is no longer visible. In a typical test of working memory, a chimpanzee is shown a toy or piece of fruit. An opaque screen then descends, blocking the object from view. Moments later the screen rises, revealing several different objects. To correctly select the previously encountered object, the animal must retain a kind of mental picture of it. Neuroscientists call this ability working memory, and it depends on the normal functioning of the frontal and prefrontal areas. If these areas are destroyed, the animal loses its ability to remember an object that is no longer in direct sight. "Out of sight, out of mind," sums up the failures that result from lapses in working memory.

Brain-imaging studies carried out during tests of working memory show sustained activity in the prefrontal cortex when something previously seen is temporarily out of sight but remains very much in mind. Additional elaborations on the content of working memory also involve activating the prefrontal areas. For example, if a volunteer undergoing a PET study tries to mentally rotate a remembered object in space, the prefrontal areas concerned with

> **A great deal of our thinking proceeds without conscious awareness.**
> —Stuart Hampshire

monitoring and manipulating the contents of an individual's working memory come into play.

Remember the anterior cingulate's pivotal role in attention? This structure works in synchrony with the prefrontal areas to help us select among competing behavioral choices, correct mistakes, and regulate emotions. Patients with disturbances in the prefrontal or anterior cingulate have great difficulty doing some or all of these things.

As children, our capacity for consciously exerted emotional self-control depends on the healthy maturation and functioning of the anterior cingulate. As we grow older, linkages between the cingulate and the later developing prefrontal areas enhance this capacity. Many criminals and psychopaths suffer from disturbances in the prefrontal-cingulate axis, which helps explain why they have trouble restraining their emotions and making wise decisions about the consequences of their behavior.

The anterior cingulate and prefrontal cortex also play important roles in obsessive-compulsive disorder. Persons afflicted with this disorder are over-controlled and their emotional reactions are out of whack. Psychosurgical operations to remedy this condition often involve the deliberate creation of lesions in parts of the brain connected to the prefrontal cortex and the anterior cingulate.

But neither of these key areas, working alone or in concert, are responsible for consciousness. All attempts to locate a single "seat of consciousness" have failed. In the words of neuroscientist T. E. Martin, who has used fMRI (functional magnetic resonance imaging) to study consciousness in normal volunteers, "The functions of attention, working memory, and sensorimotor coordination are not located in a single, discrete brain area.… complex mental operations rely on the coordinated activity of widely distributed brain regions."

Indeed, research reveals that consciousness is based on the operations of many discrete brain areas, referred to as modules. Of course, this raises the question, how and where are all these modules integrated into the unified whole we call consciousness? And how does all this separate activity become integrated into *my* consciousness and *your* consciousness? Philosophers refer to this as the homunculus problem, because some early microscopists thought they could detect a tiny person—what they called a homunculus—in human sperm or the fertilized egg of a human. Where in the brain, they ask, is hidden the tiny man or woman who makes sense out of all the brain's widely distributed activity? Even the most modern imaging techniques have failed to reveal a homunculus.

Four decades of research, however, indicates that the left hemisphere is more involved than the right hemisphere in terms of conscious experience. In the 1950s and 1960s, Roger Sperry earned a Nobel Prize for being the first to probe the mental experiences of what are known as split-brain patients—people who have undergone surgical

■ FOLLOWING PAGES: Mentally departing his prison cell for a peaceful spot within his mind, inmate Charles Peacher meditates with Brother Rasa, a Hindu minister. "Anger is probably the dominant emotion in the prison," maintains Rasa. "Meditation lets the men get that anger under control."

severing of the corpus callosum, the bridge of fibers that connects the right and left hemispheres and allows them to "cross-talk" with each other. This operation is performed on some epileptic patients as a last-ditch effort to prevent the spread of seizure discharges from one hemisphere to the other.

Sperry and others discovered that language is mediated primarily in the left hemisphere. Since language plays such a dominant role in our conscious lives, the left hemisphere dominates consciousness. For example, if I ask, "What are you thinking right now?" you will use words to describe your present state of consciousness. That answer, being verbal, depends upon the activation of your left hemisphere.

Split-brain research has provided neurologists with a number of intriguing insights on the workings of the human mind. Sperry's associate, Michael Gazzaniga, carried out experiments in which two pictures were presented simultaneously to a split-brain patient, one directed exclusively to the right hemisphere (shown to the left eye only), the other exclusively to the left hemisphere (via the right eye). The patient was then asked to choose from among several other pictures those that were most closely associated with the two pictures first shown. An example: A picture of a chicken claw is flashed to the left hemisphere, and a picture of a snow scene to the right. When asked to make his selections, the subject points with his right hand (controlled by the left hemisphere) to a picture of a chicken and with his left hand (controlled by his right hemisphere) to a picture of a shovel. At this point, Gazzaniga asks "Why did you do that?" The patient responds "Oh, that's easy. The chicken claw goes with the chicken and you need a shovel to clean out the chicken shed." What's going on here?

Gazzaniga's question, being verbal, activated the subject's language-mediating left hemisphere to produce an answer. But since the left hemisphere was privy only to the drawing of the claw and knew nothing of the snow scene seen by the right hemisphere, it did the best it could under the circumstances. It came up with a reasonable explanation as to why the right hand was pointing to a chicken while the left hand was pointing to a shovel: A shovel is a perfectly reasonable thing to be using when cleaning out a chicken shed. The context of a shovel as part of a snow scene did not occur to the subject, since the snowy image had never entered his conscious awareness.

"In other words, the left brain, observing the left hand's response, interpreted the response in a context consistent with its own sphere of knowledge—one that does not include information about the snow scene presented to the other side of the brain," says Gazzaniga.

Stated slightly differently, the talking (left) hemisphere determines consciousness. Since the snow scene flashed to the right hemisphere never entered conscious awareness, the subject does not include it as a part of his explanation.

———————————

In another split-brain experiment, the command "walk" is flashed exclusively to the non-speaking (right) hemisphere. The patient gets up and starts to leave the testing area. When asked where he is going, he might say something like, "I'm going to get a Coke." He doesn't mention the "walk" command—because his language-mediating left hemisphere never saw it, and therefore he is not consciously aware of it. He walks because the command is

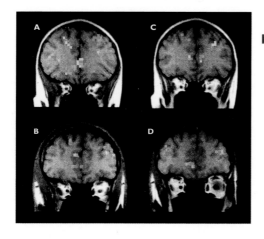

received and processed by his right hemisphere. His verbal response is merely an attempt at a plausible reason for his action, one that is totally unrelated to the actual cause (the flashed command to walk).

For this to occur, the right hemisphere obviously must have some rudimentary ability to read simple words such as "walk." But this reading is never conveyed to the left hemisphere, nor is it accompanied by conscious intent. The person simply finds himself carrying out an action directed by the "unconscious" right hemisphere, without any idea why. To cope with this dismaying situation, the left hemisphere invents an explanation. Explains Michael Gazzaniga, "The left brain is constantly constructing causal relations among elementary events that transpire inside and outside our heads."

Gazzaniga calls the brain's left hemisphere the Interpreter, defining it as "a device that seeks explanations for events and emotional experiences." In another experiment, he projects an emotionally arousing film clip to the right hemisphere of a female split-brain patient. In the clip, one man pushes another off a balcony and then tosses a firebomb at him. After showing this movie to the woman's right hemisphere, he asks her what, if anything, she had just seen. Her response is something like, "I don't really know what I saw. I think I just saw a white flash."

But when asked how she feels emotionally, she says, "I don't really know why, but I'm kind of scared. I feel jumpy. I think maybe I don't like this room, or maybe it's you; you're getting me nervous."

In this instance, the right hemisphere responds to the film clip. The woman's left hemisphere (which didn't see the clip because of the disconnection between the two hemispheres) doesn't know why

she suddenly feels anxious. But in answering the question, the woman activates her left hemisphere's Interpreter, which comes up with a plausible albeit false explanation for her mood change—that Gazzaniga is making her nervous.

"This kind of effect is common to all of us," says Gazzaniga. Indeed, most of us with normal connections between the left and right hemispheres can recognize similar situations in our mental lives. Throughout the day we're constantly carrying on a kind of inner dialogue, self-talking about how we should respond to the people and events around us. This inner dialogue comes closest to capturing the essence of consciousness.

Novelists like James Joyce and Virginia Woolf captured in their fictional masterpieces *Ulysses* and *Mrs. Dalloway* the intimate, almost inseparable linkage between inner language and consciousness. Their characters' moment-to-moment thoughts pour onto the page in a "stream of consciousness" technique, in which consciousness and language become one. Indeed, it's the absence of language that leads many philosophers to conclude that animals can't be conscious, since they haven't evolved a language sophisticated enough to enable self-talk. Yet the deception experiments mentioned earlier show that primates seem capable at least of awareness, if not consciousness. All of which brings us back to basic definitions. When all is said and done, what is consciousness?

In *The International Dictionary of Psychology*, British psychologist Stuart Sutherland suggests a suitably informal yet encompassing definition that I think captures the subtleties:

Dome-headed stock traders in Tokyo take a virtual vacation, soaking up soothing music to improve their mental attitudes and reduce the ill effects of stress. Such therapy has been shown to help the brain modify its response from "fight-or-flight" to "relax-and-smell-the-flowers."

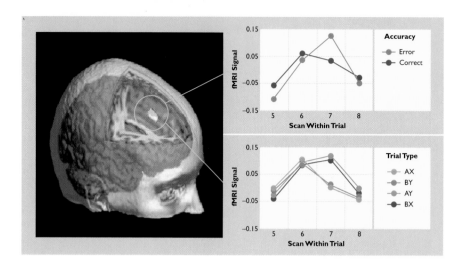

Focusing on the anterior cingulate—an area devoted to error prevention—this fMRI indicates maximum activity when the person is uncertain or knows he is wrong. Such findings support the theory that the brain uses conflict as an alerting signal in order to regulate its own cognitive functions and adapt to changing demands from the outside world.

"The having of perceptions, thoughts, and feelings; awareness. The term is impossible to define except in terms that are unintelligible without a grasp of what consciousness means....Consciousness is a fascinating but elusive phenomenon; it is impossible to specify what it is, what it does, or why it evolved."

One of the difficulties here springs from the requirement that we somehow separate consciousness out from the rest of the operations of the mind. Neuropsychologist Richard Gregory suggests a useful metaphor: "Imagine consciousness as a beam of light of a torch or searchlight, directed around the dark universe of the Mind. We are the beam of light, we are an eye looking along the beam, and we see nothing outside. But there is a great deal around it, and much that is never illuminated. All this is Mind. Present consciousness is what is lit by the patch of light of the beam."

Gregory adds that we could be "easily fooled into thinking that consciousness is the whole of Mind—for we only experience what is in the beam." Not only that, but mental operations outside of awareness may influence the mind—as shown by the split-brain and masked-priming experiments.

Perhaps our difficulties in coming to terms with consciousness stem from our insistence on the mistaken notion that consciousness is somehow a *thing* rather than a process. MIT linguist Noam Chomsky compares consciousness to gravity, electromagnetism, and other processes and properties of nature. "The simple act of a walnut rolling off a roof and onto the ground illustrates action at a distance—in this case gravity. Nobody asks 'Where is the gravity that caused the walnut to fall to the ground.' Such a question is nonsense because gravity is not a thing but an invisible elementary property of nature. Perhaps consciousness is best thought of as an emergent property of brains that have reached a critical level of complexity."

Samuel Johnson also addressed the mysterious nature of consciousness and why we may need new ways of thinking about it: "Matter can differ from matter only in form, bulk, density, motion, and direction of motion: to which of these, however varied or combined, can consciousness be annexed? To be round or square, to be solid or fluid, to be great or little, to be moved slowly or swiftly one way or another, are modes of material existence, all equally alien from the nature of cognition."

Any new ideas we may come up with in regard to consciousness can't help but influence how we think about the relationship of mind to brain. The so-called mind-body problem (better thought of as the mind-brain problem) boils down to how well we relate our thinking, remembering, and other cognitive processes to the brain. As we have seen with PET scans, correlations aren't hard to find. But a correlation doesn't provide an explanation nor does it further our understanding beyond a certain point.

For instance, the movement of my fingers on the keyboard as I write these words can be correlated to changes in the prefrontal, motor, and premotor areas of my brain—as revealed through changes in PET scans. But this correlation leaves the really important questions unanswered. Hidden between the scan images and my conscious awareness of typing specific words lies what philosophers call an "explanatory gap." As I type, I experience consciousness as John Locke described it—"the perception of what passes in a man's own mind." The PET images, despite their monitoring of chemical and electrical changes, can never define the events occurring in my mind. And to speak of a "consciousness" of chemical and electrical events is to talk nonsense. In fact, such perception would no doubt prove a hindrance rather than an aid to understanding consciousness.

The brain functions on many different levels. While it would be nice to be able to correlate each to the next, this may not always be possible. Indeed, does it even make sense to ask, "What is happening in my brain's potassium channels or serotonin receptors when I decide to contact an online travel service and book a flight tomorrow to Geneva?" I don't think so. To ask such a question is to be guilty of what philosopher Gilbert Ryle called the category mistake. This refers to our tendency to lump different things together instead of carefully distinguishing them. Both apples and oranges are fruits but, according to a traditional caution, we shouldn't mix the two when making a point. Only compare things that are in the same category.

Analogies provide a useful means of comparison without mixing apples and oranges. My favorite one compares brain and consciousness to a clock and time. The position of the clock hands (the brain) bears a constant relationship to the time (consciousness) indicated by the clock. The clock hands, like the brain, are physical and configured in special spatial positions.

Time, however, is not spatial; it lacks position or physical extension. So it is with consciousness. Although we measure time in terms of the physical position of the hands on the clock face, time cannot always be meaningfully understood in this manner. Even a clock that has stopped, for example, accurately indicates the time twice a day (once a day for 24-hour clocks).

There are limits to this analogy, of course. The clock (or brain) tells the time and can be read by any number of observers. But consciousness is a personal affair; only you are privy to it. This is a fundamental and probably unbridgeable distinction. As Colin McGinn writes in *The Mysterious Flame*, "Consciousness indubitably exists, and it is connected to the brain in some intelligible way, but the nature of this connection necessarily eludes us."

——————————

Perhaps we should stop looking for connections and locations and, instead, search for dynamic systems that move through the brain like waves through the ocean. Rodolfo Llinás, of New York University, has measured such a wave: Every 12.5 milliseconds, this powerful 40-cycles-per-second wave occurs between the thalamus and the cortex. He believes it provides an answer to what he calls the "binding problem": how the information provided by all our senses gets tied into a single, coherent whole.

Because nerve nets with billions of neurons would have impossible problems in functioning if every neuron "spoke" at the same time, neuronal activity (which is partly electrical) is organized so that important messages occur with a particular electrical rhythm. Neurons that have the same rhythm at the same time connect with each other more easily than when they have different rhythms (not unlike what happens when people dance, Llinás points out).

At a recent brain conference, Llinás used a video of the Irish step-dancers in *Riverdance*— moving together in perfect unison—to summarize how neuronal connectivity and rhythm can produce the functional state we know as the mind.

In *The Missing Moment: How the Unconscious Shapes Modern Science*, author Robert Pollack states, "The conductor in charge of bringing the symphony of consciousness out of the brain's separate centers is a synchronizing wave of electrical activity that sweeps regularly through the brain from behind the forehead to behind the nape, forty times per second." He adds, "This wave links the centers responsible for processing sensory information to one another as well as to other centers responsible for unconscious and conscious activities of the mind, in particular the amygdala, the hippocampus, and the frontal cortex, where, broadly speaking, emotional states are generated, long-term memories stored, and the intentions to speak and to act generated."

Other researchers have come up with waves of greater or lesser frequencies than have Llinás and Pollack. But the synchrony of these waves may be more important to consciousness than the frequency of their occurrence. That's because whatever information is conveyed to the brain must be integrated. The thalamus is a key area for binding together the disparate elements that make up mind and consciousness, suggests Llinás. As the stop-off point on the way to the cerebral hemispheres for every sensation except smell, he adds, the thalamus is in a good position to act as "a hub from which any site in the cortex can communicate with any other."

A slightly different theory, the dynamic core hypothesis advocated by Nobel Prize winner Gerald Edelman, is based on the idea that some as yet undefined system or wave moves through the brain and activates specific circuits that create conscious awareness.

Will an emphasis on waves provide a better understanding of consciousness? Certainly forces and energy fields have become central to our understanding of the physical world. According to contemporary physics, "solid" matter isn't so solid after all, but is composed of atoms separated by large empty spaces. Why not think of consciousness in similar terms? In the words of Oxford philosopher Galen Strawson, "It can seem natural to think of consciousness as a form or manifestation of energy, as a kind of force, and even perhaps a kind of field."

So our search for an explanation of consciousness ends not with certainty, rather with inconclusive yet intriguing possibilities. To experience consciousness we have only to direct our attention inward to our own thoughts. But we can't come up with a wholly satisfactory explanation of that consciousness, any more than St. Augustine could explain time. "*Si non rogas intelligo*," he wrote, roughly translated as, "The more I set myself to think of it, the less I understand it." Though we are all conscious, no one has yet explained how or why consciousness is even possible.

No wonder philosopher Arthur Schopenhauer referred to the problem of human consciousness as "the world knot."

■ After marathon praying and chanting, the head of a Namibian faith healer briefly emits an eerie flash of light, according to some witnesses. Scientists and theologians continue to debate whether reported phenomena such as this are due to supernatural or physiological causes.

SLEEP AND DREAMING

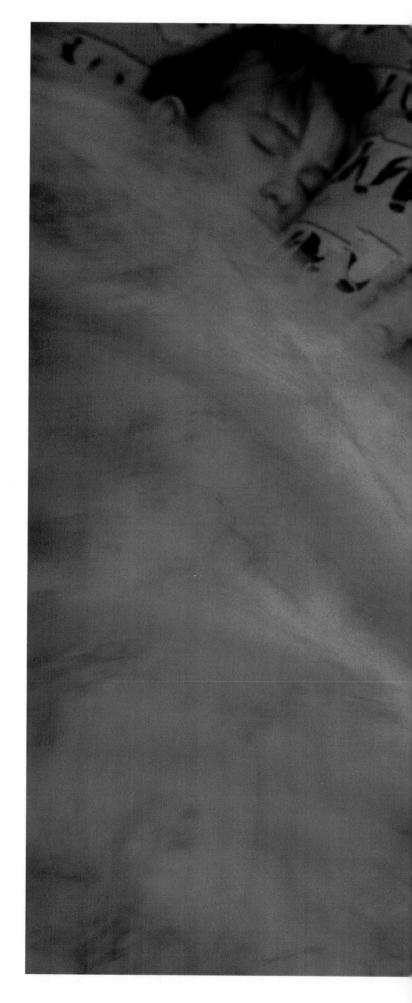

We sleep away a third of our lives. This holds true whether we are Australian aborigines curling up beneath the stars, or urban dwellers slipping between satin sheets in high-rise apartments. As Homer phrased it in the *Iliad*, "Men and gods alike [bow] to sleep in submission."

Ancient Greeks attributed sleep to the influence of the winged god Hypnos, brother of Thanatos, the god of death. Later Greeks concentrated less on gods and more on naturalistic explanations such as the four elements (fire, water, air, and earth) and their associated humors. In a peculiar twist, the philosopher Alcmaeon suggested that sleep stemmed from a temporary loss of blood to the brain; waking occurred once that blood was restored. Its failure to return at all resulted in death.

Others emphasized the restorative powers of sleep, believing that it provided the body a needed interval for self-repair. Shakespeare incorporated this concept in a soliloquy by Macbeth:

> *Sleep that knits up the ravelled sleeve of care,*
> *The death of each day's life, sore labour's bath,*
> *Balm of hurt minds, great nature's second course,*
> *Chief nourisher in life's feast*
> (*Macbeth*, c. 1606, Act II, scene II)

Scientifically speaking, however, the first real

■ Same bed, different worlds: Although these three children sleep together, the brain of each follows its own dynamic rhythms. Electrical and chemical activity, dream content, and duration of various sleep stages differ for every individual.

> **Balm of hurt minds, great nature's second course, Chief nourisher in life's feast.**
> —William Shakespeare

breakthrough in our understanding of sleep took place more than 300 years later. In 1953 Eugene Aserinsky, an assistant in a sleep laboratory at the University of Chicago, noticed eye movements in sleeping subjects. Curious, Aserinsky hooked up his eight-year-old son to an electroencephalogram (EEG) machine and studied the sleeping child's eyes. To his surprise, he noted that the EEG recorded during his son's eye movements looked similar to the record made during wakefulness. This seemed inconsistent with the idea that sleep was a quiescent, reparative state. What could explain such a strangely contradictory finding?

Aserinsky tried waking his son at different times, and found that whenever the boy was awakened during an episode of what became known as REM sleep (for rapid eye movement), he would report having had vivid dreams. When awakened at other times, he never recalled any dreams.

Subsequent experiments by Aserinsky and others confirmed the REM connection to dreaming. Between 80 and 95 percent of sleepers awakened during REM sleep reported dreams, compared to only 7 per cent of those awakened from non-REM (NREM) sleep.

Researchers then decided to monitor the brain waves of volunteers during an entire night's sleep. They found that normal sleepers go through regular cycles of REM sleep alternating with four levels of NREM. About 60 to 80 minutes after first falling asleep, the brain's activity begins to increase; the sleeper drifts from deep Stage IV sleep to Stage III, Stage II, and then Stage I—at which time rapid eye movements begin. After about ten minutes in REM, the brain drifts back down through the four stages. The process is repeated every 90 to 100

minutes, but as night progresses, the REM periods lengthen and the NREM periods shorten. This predominance of REM late in the sleep cycle explains why most dreams occur in early morning. In fact, if you want to increase your chances of dreaming and remembering those dreams, set your alarm to go off an hour or two before you need to get up and, when awakened, turn the alarm off and go back to sleep.

In all, REM sleep occupies 20 to 25 percent of the sleep cycle; the rest is NREM sleep. Thus in a typical eight-hour sleep period, about two hours will be spent in REM. Assuming an average life span of 70 years, you can expect to spend about 50,000 hours—six years—dreaming.

As sleep researcher J. Allan Hobson notes in his book, *The Dreaming Brain*, "The fact that we spend so much time in REM would seem to indicate that evolution has placed a high priority on putting the brain into this state; something profoundly important to our biology must be going on for nature to devote so much time to this."

Although sleep is commonly associated with rest, REM sleep actually represents a state of brain and body activation. In addition to rapid eye movements, the heart beats faster and the breathing rate increases. Yet the sleeper appears tranquil almost to the point of paralysis, despite vivid dreams that can involve intense physical and emotional responses on the part of the dreamer. What accounts for this paradox of increased brain activity occurring during sleep, a state characterized by the outward appearance of total inactivity? The answer emerged from a series of experiments carried out on cats.

The cat is a good subject for sleep research, not only because it experiences REM sleep, but also because it sleeps about 50 percent of the time, a pattern it shares with its jungle-predator cousins. In the late 1950s, sleep researchers François Michel and Michel Jouvet, working in Lyon, France, made a surprising find. In some cats they cut the nerve fibers that connect the pons to the limb muscles—and found that when in REM, the animals would hiss, stand up, pounce, or exhibit other "hallucinatory behaviors," apparently in response to whatever they encountered in their dreams. Such behavior did not occur in cats that had not undergone the severance of the fibers, suggesting that REM sleep actively inhibits attack and defense postures that would be perfectly normal in the waking state.

———————————

How do we know that these "hallucinatory behaviors" were in response to dreams? Mostly because we experience a similar state of REM-related "paralysis" associated with frightening dreams. Those who suffer from a rare illness called REM sleep behavior disorder fail to become "paralyzed" during REM sleep, and physically respond to their dreams. One of my patients once fractured his nose when, in order to escape from a dreamed pursuer, he leaped out of bed and charged into the bedroom wall. After waking on the floor he recalled the chase and his attempts to outrun his pursuer. Treatment with a drug called clonazepam strengthened the functional connection between his brain stem and muscles, thus reestablishing the temporary muscular paralysis that normally accompanies REM sleep.

Michel and Jouvet also discovered a control system in the pons that is the source of both the EEG activation and the rapid eye movements observed during REM. In synchrony with each

REM episode, periodic signals known as pontine-geniculate-occipital waves (PGO) ascend upward from the pons to the lateral geniculus (a way station on the neural pathway involving vision) then on to the visual cortex (the occipital lobe). During REM sleep, widespread excitation occurs in individual neurons along the PGO route.

"REM sleep is a period of both global and specific changes in the activation of neurons and flow of information throughout the brain," observes Hobson. During early NREM stages the brain is taken "off-line"; it stops processing information about the outside world. As sleep progresses into REM, the process of disengagement from the outside world continues. Dreams and other inner representations become the controlling forces.

Changes also occur at the chemical level. Sleep and wakefulness are marked by periodic alterations in a delicate balance involving three neurotransmitters within the brain stem. On one side of this balance beam are the noradrenaline and serotonin centers, which contain pacemakers that fire spontaneously during active physical and mental activities, releasing these two neurotransmitters.

"When we're awake," observes Hobson, "the brain cells that produce norepinephrine and serotonin are active, and we can line up our thoughts, think logically, and process external data." In contrast, neurotransmitter output falls off in these centers during states of relaxed wakefulness, when nothing much is happening, and during the drowsy episodes that just precede the onset of sleep.

On the other side of the balance beam are the neurotransmitter acetylcholine and its centers, which contain no pacemakers. The cells of these centers tend to be generally quiescent during

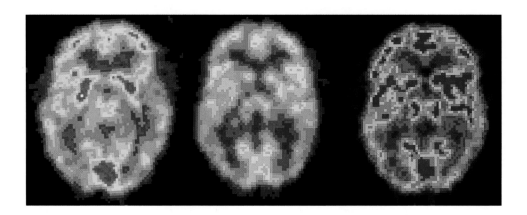

wakefulness, perking up only when something special happens that captures attention.

"Thus the waking brain is bathed in constant levels of norepinephrine and serotonin and receives pulsatile boosts of these two chemicals under routine conditions and acetylcholine under novel conditions," says Hobson.

As sleep ensues, the modulatory neurons become less active; as NREM sleep deepens, the balance of neurotransmitters changes. Still later, at the onset of REM sleep, the balance has shifted even further: Norepinephrine and serotonin are essentially shut down, while the acetylcholine neurons are fully active.

In essence, specific groups of neurons in the brain stem regulate sleep and wakefulness. Referred to collectively as the reticular core, these neuronal clusters send elaborate projections upward to the hypothalamus, midbrain, and cerebral cortex, and downward to the spinal cord. Thanks to the wonderfully coordinated action of the pacemakers and the three neurotransmitters, sleep and waking follow a regular pattern that differs from one person to another. Of course, the amount of sleep required also differs considerably among individuals, although the physiological mechanisms of sleep remain the same. Why is this so? And how much sleep do we really need?

The 12th-century Rabbi Moses ben Maimon, known to history as Maimonides, urged all his followers to sleep for one-third of the day, or eight hours. The happy citizens of Sir Thomas More's Utopia also slept for precisely eight hours. Such historical and literary sources may help explain the common belief that eight hours of sleep is the gold standard for which we all should strive.

But what of people like Miss M., a retired nurse in London, who slept for no more than an hour every day and considered additional sleep—in herself or others—"a waste of time." Researchers who documented her "healthy insomnia" in a sleep laboratory concluded, "We are very much at a loss to explain why Miss M. should be as she is."

Actually, Miss M. and many other individuals represent fascinating exceptions to a basic rule. In the general population, sleep requirements are distributed along a classic bell-shaped curve that ranges from about 4.5 to 10.5 hours and peaks between 6.5 and 8.5 hours. The length of sleep required does not reliably correlate with intelligence or personality; while Thomas Edison and Napoleon Bonaparte were both short sleepers, Einstein was notorious for sleeping longer periods than his less gifted colleagues.

———————

And yet sleep serves a necessary function in all our lives. A classic experiment in sleep deprivation took place in 1959, when radio personality Peter Tripp entered a transparent broadcasting booth set up in Times Square to begin a marathon of sorts; he planned to remain awake for the next 200 hours.

For the first few days, things progressed reasonably well. But then Tripp began experiencing hallucinations—insects were marching across his studio console; a rabbit was sitting across from him by the window. By the 100th hour of wakefulness his hallucinations grew grimmer: Tripp believed he saw a man wearing a suit made of crawling worms, and he began to complain that somebody was putting drugs in his food in order to make him fall asleep. Shortly after passing his 200-hour

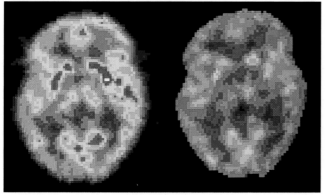

Sleep's diverse states of mind take on distinctly colored patterns in PET scans, which use radioactive tracers to measure metabolic rates and blood flow that accompany various mental activities. Stages shown depict (from left) wakefulness, non-REM sleep, REM sleep, normal sleep, and sleep deprivation.

goal, Tripp thought the doctor called in to examine him was actually an undertaker, and ran out into the street screaming for help. At this point his self-styled experiment was halted; Tripp was taken to a hotel room, where he slipped into a 13-hour-long sleep. When he awoke, the hallucinations and delusions had disappeared. But he would experience several weeks of mild depression before his usual outgoing personality returned.

Another wakefulness marathoner, 17-year-old Randy Gardner, set a record of 264 sleepless hours. He also expressed some strange sentiments and mental lapses. At one point he announced that he was a famous black professional football player. (He was clearly Caucasian.) Later he claimed that a radio-show host was trying to make him look foolish. After sleep and a brief recovery, Gardner also returned pretty much to normal.

Peretz Lavie, head of the sleep laboratory at the Technion-Israel Institute of Technology in Haifa, Israel, accepts that Randy Gardner's feat proves that "people who are absolutely determined are able to continue without sleep for several days, especially when they have the support of constant monitoring and excitatory stimuli."

But the experiences of Gardner, Tripp, and others also confirm that prolonged sleeplessness is an unpleasant and even intolerable experience that, if carried to extremes, ultimately leads to serious mental disturbance. Fortunately such episodes of what's called sleep deprivation psychosis occur in only a small number of people pushed to those extremes. Usually, the symptoms disappear once the person has slept.

Even for those of us not interested in setting wakefulness records, it's important to regulate our lives so that we get proper amounts of sleep. Like the ancients, most modern sleep researchers agree that sleep serves a restorative function.

"It has been difficult to move beyond the subjectively compelling but scientifically unsatisfactory notion of sleep as [a form of] rest," writes Hobson, pointing out that sleep deprivation experiments on rats resulted in the death of all subjects after 4 to 6 weeks. The animals could not maintain a steady weight or constant body temperature. Their ability to resist infection was seriously impaired. Sleep, it seems, is an energy- and heat-conserving response that normally takes place at night, when ambient temperatures are usually at their low.

Sleep also functions as a consolidator of information. We spend more time in REM sleep during periods when we are learning new things; if REM sleep is interrupted, we remember less the next day. Hobson speculates that the dominant action of acetylcholine during sleep is to help consolidate memories already in the system. The phrase "already in the system" is crucial, however. Numerous studies have debunked the popular myth that you can learn a foreign language simply by leaving language tapes playing as you fall off to sleep.

"The many functions of sleep are still being characterized, and it is clear that sleep is not a

FOLLOWING PAGES: Normal sleepers— even those who claim they never budge once they fall asleep—average eight to twelve posture shifts during the night, as this time-lapse sequence shows.

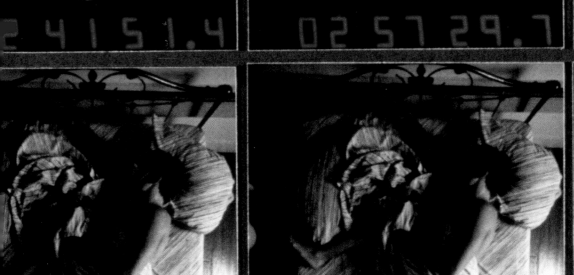

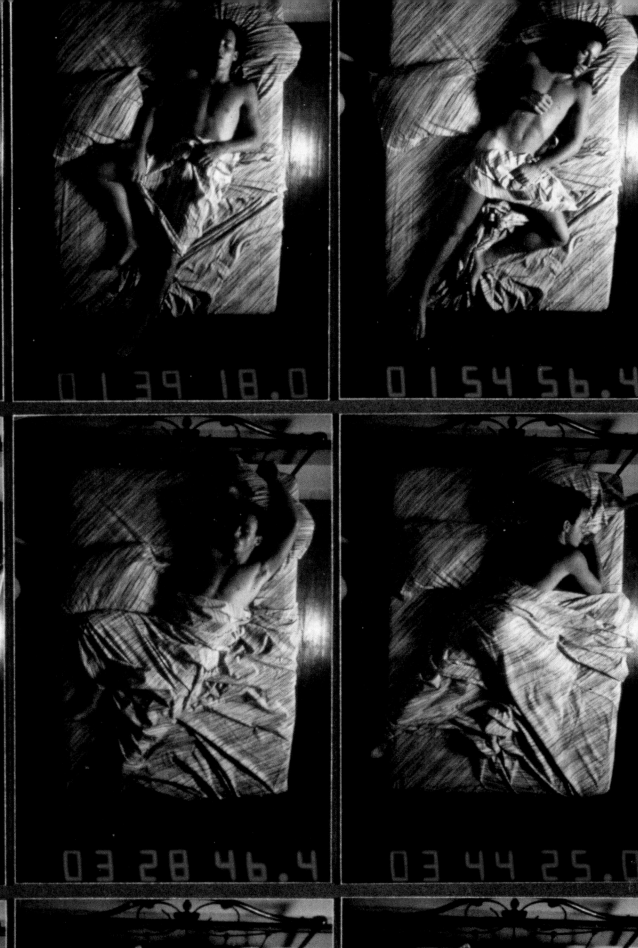

01 39 18.0 01 54 56.4

03 28 46.4 03 44 25.0

passive state but a dynamic condition during which memories are consolidated, restorative hormones are released, and neuronal excitability is modulated," adds Beth A. Malow, of the Michael S. Aldrich Sleep Disorders Laboratory, at the University of Michigan medical school.

This much is certain: We are endowed with a biological clock designed to optimize our physical and mental performances during daytime hours. The natural pattern includes work during the day, recreation during the evening, and sleep at night.

Yet nationwide, sleep patterns are increasingly deviating from our natural rhythms. We are moving towards a 24-hour, 7-day-per-week economy. Less than a third of employed Americans work a standard work week, defined as 35 to 40 hours on a fixed daytime schedule, Monday through Friday. In fact, only a slim majority—54.4 per cent—regularly work a fixed schedule of any sort, defined as the same hours every day during a five-day week. Nonstandard work schedules are becoming the norm—as are the increasingly apparent consequences.

"Shift work is a remarkably efficient device for disrupting an individual's normal sleep-wakefulness cycle," states Stanley Coren, author of *Sleep Thieves*. So efficient, in fact, that 20 percent of workers find shift work so disruptive to their physical and mental health that they voluntarily give up their jobs. But the quantity and quality of our sleep has consequences that extend far beyond employment.

Shift work and other nonstandard schedules often lead to sleep disturbances, gastrointestinal disorders, chronic fatigue, lack of energy, and even depression. Not surprisingly, such disturbances place a severe burden on families. Among couples with children and married less than five years, the likelihood of separation or divorce increases sixfold if the man works nights instead of days. When women who have been married more than five years work at night, the odds of divorce or separation soar three times higher.

"Moreover,…the increased tendency for divorce is not because spouses in troubled marriages are more likely to opt for night work; the causality seems in the opposite direction," according to Harriet B. Presser, a sociologist and author of *Toward a 24-Hour Economy*.

———

Timing one's sleep is also important. Major low points in human alertness occur between 1 and 4 a.m. and between 1 and 4 p.m. Most single-vehicle accidents also occur during these six hours. Even when normally rested, people are most prone to drowsiness, fatigue, and inefficiency at these times, a fact that lends rationale to taking a daily siesta during the afternoon low point.

Increased accidents of all sorts also occur during the spring shift to daylight savings time, when we lose a precious hour of sleep. Traffic accidents rise by about 7 percent, and the chance of death from injuries sustained in such accidents increases by 6 percent. There is, of course, another way to look at the same data: An extra hour of sleep can reduce highway accidents by 7 percent and accidental deaths by 6 percent. Fortunately the daylight-savings shift occurs only once per year, and most everyone adjusts after the first day or so.

Among the sleep deprived, the potential for accidents is even worse. Both the 1979 nuclear incident at Pennsylvania's Three Mile Island and the 1989 oil spill of the *Exxon Valdez* involved sleep-deprived people making serious errors during the early morning hours.

"All of this might have been avoided if the operations staff had had enough sleep to allow them to react in a quick and intelligent manner," maintains Stanley Coren. As novelist Aldous Huxley puts it, "That we are not much sicker and much madder than we are is due exclusively to that most blessed and blessing of all natural graces, sleep."

One way to buck the national trend toward chronic sleep deprivation is the afternoon nap. A Congressional panel on biological rhythms (on which I was privileged to serve) concluded, "Brief naps taken during temporary periods of extended work may enhance subsequent performance, and naps taken during time off may help to offset the sleep debt typically associated with shift work."

In fact, there is good reason to believe that napping is not only a practical approach to sleep deprivation but also a natural aspect of human behavior. All of us nap as children, and give up the practice only grudgingly when we start school. More than 50 percent of college students reestablish the napping habit, indulging in an afternoon snooze at least once a week. With increasing age—particularly after retirement—many perfectly normal people tend to cycle back to their childhood pattern of taking short daily afternoon naps.

The mid-afternoon decrease in mental and physical performance probably reflects an inborn tendency to drowse. If you've ever returned to the office after a heavy lunch you can probably personally attest to feelings of sluggishness and a desire to lie down for a few minutes or at least lay your head on your desk. Should you give in to such impulses, don't feel guilty; you're only emulating the sleep

habits of no less an anti-sleep zealot than Thomas Edison, who once declared that sleeping eight hours a day was a "deplorable regression to the primitive state of the caveman."

There's an enduring story that Henry Ford once made a surprise visit to Edison's lab, only to find that the great inventor was napping. Ford, well aware of Edison's attitude towards sleep, said: "I understood that Mr. Edison didn't sleep very much." "Oh, that's true," a technician responded. "He doesn't sleep very much at all, he just naps a lot."

Other famous nappers include Winston Churchill (one to two hours between lunch and dinner), Leonardo da Vinci (fifteen minutes every four hours), Napoleon, and Salvador Dali.

While naps are certainly beneficial, it's unlikely that many of us could get along on naps alone. Italian playwright and actor Giancarlo Sbragia once tried the nap-only strategy attributed to Leonardo. He stopped going to bed at night and with the help of an alarm clock trained himself to lie down for a brief nap every four hours.

During the early stages of this six-month-long experiment, Sbragia felt exhilarated by the new experience of a twenty-two-and-a-half-hour day. But a few months later, he described himself as a "psychological wreck"; his artistic work began to deteriorate. After suffering "a kind of imaginative damage," he terminated the self-imposed experiment and resumed a regular, eight-hour sleep session. Interestingly, Sbragia observed of his return to normalcy, "I recovered my dreams."

Sleeping and dreaming; It's so natural to link the two that the Spanish language doesn't even make a distinction between them. Novelist Martin Amis wryly comments, "You're always asleep when [dreams] happen. Perhaps sleep is to blame, pulling the wool over our eyes in that deceitful way it has. Dreams wouldn't dare do what they do to me when I'm awake. That's why they wait until I'm asleep before they do it."

Amis, of course, is not the first to puzzle over the relationship of sleeping to dreaming, and the overall purpose served by making nightly forays into a bizarre and phantasmagoric dream world. The ancient Egyptians believed that dreams were a way to see the future and communicate with the gods. They even journeyed to special places and relied on trained priests to interpret their dreams. Prophetic dreams are also mentioned throughout both the Old and the New Testaments.

When the Roman emperor Nero related an especially terrifying dream, he was told: "It is neither the gods nor divine commandments that send the dreams down from the heavens, but each of us makes them for himself." The orator Cicero also undercut the importance of dreams, by questioning why the gods would choose to communicate with people through such unreliable means.

The insightful Greek physician Hippocrates first suggested that the brain was the origin of dreams, a point borne out by the ensuing two thousand years of dream research. But if dreams originate in the brain, are they influenced by our experiences while we're asleep, or are they solely the products of internal brain processes?

In hopes of answering this question, Louis Maury, a 19th-century researcher, had assistants carry out experiments on him after he fell asleep. They opened perfume bottles under his nose,

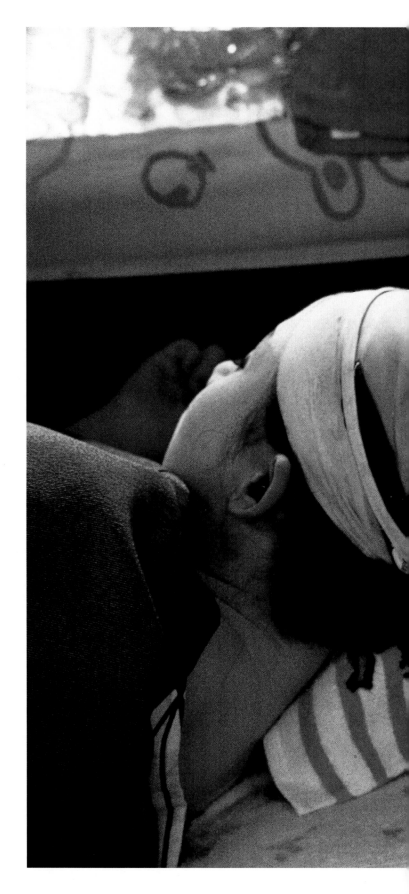

Geishas slumber on cloth-draped ceramic pillows, traditionally believed to properly cradle the head and align the spine, thus ensuring comfort and optimal flow of the life force.

dropped water on him, and made various noises, to see if their actions might influence his dreams.

Maury theorized that dreams result from physiological changes in the brain, brought about by impressions received from the sensory organs while we sleep. The results of his experiments seemed to support that contention: A burning match held to his nose while asleep in a room with an open window gave rise to a dream of a ship, an explosion, and then drifting on the ocean.

Charles Dickens later incorporated Maury's dream theory into *A Christmas Carol*. Scrooge at first attributes his visions "to a slight disorder of the stomach." He tells the Ghost, " You may be an undigested bit of mutton, a blot of mustard, a crumb of cheese, a fragment of an undone potato."

But not all dreams are related to immediate internal or external events. We often dream of people and places from our distant past. In 1899 Sigmund Freud attempted to explain such occurrences by suggesting that the remembered part of a dream was only the tip of a psychic iceberg. Deeper down, he believed, was what he called the dream's latent content, its symbolic meaning.

Freud believed that sex, death, and violence were too traumatic for the rational conscious mind to comfortably deal with. Such material was therefore relegated to the unconscious and typically became disguised in symbolic representations. The result was a "forbidden" gratification that carried with it the seeds of neurosis. Treatment consisted of interpreting the disguised latent content and thereby reconciling the conscious with the unconscious.

In fact, the idea that dreams often reflect unconscious wishes that elude the strict censors of our conscious mind is not new. Plato wrote of

desires that "bestir themselves in dreams when the gentler part of the soul slumbers and the control of reason is withdrawn; then the wild beast in us, full fed with meat or drink, becomes rampant and shakes off sleep to go in quest of what will gratify its own instincts."

Today, of course, Freudian psychoanalysis exerts far less influence on dream theory than it once did. Freudian interpretations of dreams do not possess the precision and verification that are prerequisites for a science; given the same patient and the same dream, two different Freudian analysts are likely to produce different interpretations. While this doesn't prove that dreams lack meaning, it also serves as a caution about putting stock in any theory that strays too far from a biological underpinning.

At this point a summary seems in order: Dreams are the result of physicochemical changes within the brain. They can be "captured" within the sleep laboratory by waking the dreamer during a period of REM activity, which occurs about six times per night. In terms of content, dreams incorporate both recent and past experiences. Dream content typically includes the bizarre, the frightening, and the physically impossible, such as dreams of flying. Given all this, are dreams meaningful? Can they be interpreted intelligently? How are they produced, and what brain mechanisms are involved?

For many modern neuroscientists, dreaming represents a subjective awareness of the creations of the brain produced during sleep. J. Allan Hobson compares the dream to a Rorschach card—marked with strange and ambiguous blots intended to elicit the dreamer's free associations and interpretations.

Dreams are the product of our cortex's efforts to do the best it can under very difficult operating circumstances.

—J. Allan Hobson

As mentioned earlier, the brain's chemistry and electrical activity change as we go to sleep. The cells that produce norepinephrine and serotonin—the chemicals associated with logical, rational thought—turn off. As a result, the cortex goes off-line. At the same time, the cell clusters in the pons that produce acetylcholine increase their activity. REM and NREM sleep occupy the two sides of a seesaw. An excess of acetylcholine tips the balance towards REM sleep; norepinephrine and serotonin nudge the brain towards bringing REM to a halt.

With the ascendancy of acetylcholine, the way is paved for the emergence of discontinuities, incongruities, and sometimes intensely disturbing, hallucination-like episodes that characterize the dream experience. Firing patterns originating in the pons bring all this about.

The volleys of signals that begin in the pons activate cells in the brain stem that cause the eyes to move. They also activate the cerebral cortex, thereby eliciting memories and stimulating diverse areas. But because the cortex remains off-line during sleep, it cannot provide the usual, rational explanations that a wakeful consciousness can, such as, "Relax: It's only a dream." Finally, the volleys from the pons travel to the brain's emotional centers such as the amygdala—setting into motion the anxiety and fears that commonly characterize many dreams.

"Dreams are the product of our cortex's efforts to do the best it can under very difficult operating circumstances," writes Hobson. "Indeed, this is a credit to the extraordinary creative capability of the brain. As electrical signals travel through the brain, triggering memory fragments and spurious sensory input, the cortex pulls them together into stories and visual images."

Do dreams make any sense? Or are they simply meaningless fantasies undeserving of interpretation? To Nobel Prize-winner Francis Crick—who with James Watson discerned the chemical structure of DNA—dreams represent merely one's mental trash. Along with Cambridge University's Graeme Mitchison, Crick believes that dreams result from the brain's need to rid itself of undesirable connections among neurons, to basically "unlearn" information that is either no longer relevant or that interferes with the brain's normal functioning. Thus, in the Crick-Mitchison scheme of things, dreams are useful but their content is useless—and might even be harmful.

"Attempting to remember one's dreams should perhaps not be encouraged," the two warned in *Nature*, "because such remembering may help to retain patterns of thought which are better forgotten."

If Crick and Mitchison are correct, interpreting dreams makes about as much sense as interpreting the configuration of tea leaves at the bottom of a cup or the pattern of fallen leaves on a lawn. And yet each dream is unique to the dreamer; only in Gothic novels do two people ever experience exactly the same dream. Based on our brain's unique responses during sleep, we weave a narrative that sometimes makes perfect sense and other times makes no sense. In either case, any meaning that emerges is ours alone. No seer or prophet or analyst can tell us what our dreams really mean. Jonathan Swift knew that nearly 300 years ago, when he sagely advised, "Those dreams that on the silent night intrude…are mere productions of the brain, and fools consult interpreters in vain."

REALMS OF MEMORY

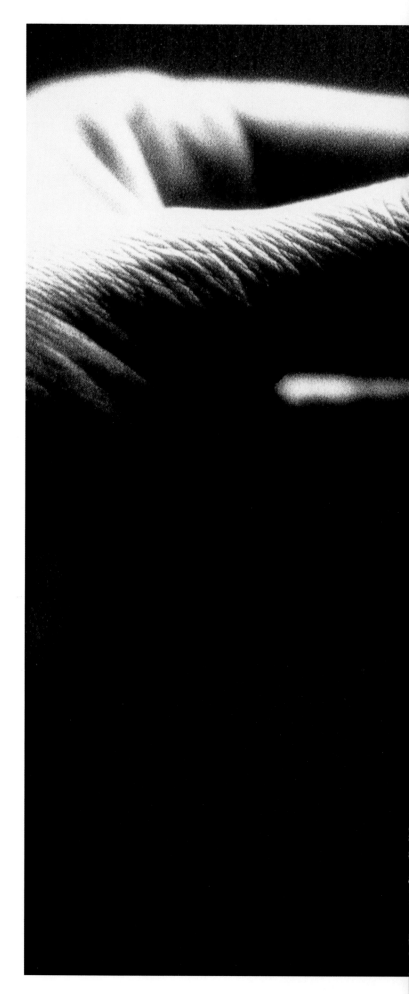

Memory enlarges our world. Without it, we would lack a sense of continuity and each morning encounter a stranger staring back from the mirror. Each day and event would exist in isolation; we could neither learn from the past nor anticipate the future. Thanks to memory, we can reach back over years and establish linkages. For example, I recently recognized a woman on a crowded street as a playground friend from my early childhood. Without memory, we would have passed each other in silence instead of celebrating our momentary reunion.

But while we recognize the value of a good memory, we also often wonder about the basic nature of the process. Is memory encoded in the brain like videotape that we can replay in a mental VCR? If so, then a person's recall of events should generally be accurate and the possibility for recalling things that didn't happen would be quite remote.

Or is memory formation a more dynamic process, one that is subject to error, distortions, even false memories? Certainly most of us can admit to occasions when we've been absolutely certain about a recollection—that turned out to be wrong. We know that confidence in our powers of recall doesn't always guarantee a correct recollection. This is true even for those people who have what psychologists call flashbulb memories—especially vivid and

■ If only it were truly photographic: Far from being a snapshot-like process, memory involves integrating complex abilities, each dependent on different mechanisms of learning and different areas of the brain.

enduring memories that involve events so shocking or surprising that they often leave pictures in the mind that the individual "will never forget."

Consider, for example, the *Challenger* space shuttle disaster of January 28, 1986. As part of a memory experiment, first-year college students were asked to recall where they were and what they were doing when they first learned about the event. Each student wrote out his or her recollection and handed it in. Four years later, the professor located the students, now seniors, and asked them the same question. Not only did their responses frequently differ from their recollections of four years earlier, but also their confidence didn't serve as a reliable measure of the accuracy of those recollections. In fact, the students who were most certain of the reliability of their recollections were often mistaken.

————————

Such findings suggest that memory is *not* like taking a snapshot at one moment and then looking at it some time in the future. Memory involves the integration of a complex collection of abilities, each of which depends on different learning mechanisms and different areas of the brain.

Imagine telling a friend about the day you surprised everybody (yourself included) by hitting a home run in a grade school baseball game. Your conversation involves several different kinds of memory. First, your memory of the game of baseball, and all the facts you can recite about it. Psychologists refer to this as *semantic memory*. Second, your memory for the specific afternoon, the uniforms, the names of the players, and your feeling of elation as you looked up and watched the ball soar into the clouds. Your recall of the specifics of that special day is called *episodic memory*. Neuropsychologists often combine these two memories into a single term— *declarative memory*—memory that we can bring into conscious awareness, reflect upon, and speak about.

But your conversation will noticeably lack another type of memory: the skill you possessed that enabled you to gauge the exact split-second to swing the bat, strike the ball, and hit the home run. This isn't something you can put into words or teach anyone how to do by verbal instruction alone. Your memory for the process lies outside of your conscious awareness and, since it can't be put into words, it's referred to as *non-declarative* memory.

Each type of memory isn't static; it changes according to time and circumstance. For one thing, your recall of the home run will undergo a gradual loss of detail with the passage of time. If you're depressed or out of sorts you will remember different aspects of that singular experience. In short, memory isn't at all like a videotape that you insert into a biological VCR within your brain; memory is a creative reconstruction of the past.

Everyday attempts to recapture our past through our memory seem to work for most of us most of the time. And while improving our memory is a common and even laudable goal, we probably wouldn't want to accept the consequences of having a perfect one. Russian neuropsychologist Alexander Luria wrote of such a person, a patient he identified as S in his book, *The Mind of a Mnemonist*.

S, he reported, could reproduce a series of words or numbers in forward or reverse order, even after a lapse of many years after the initial testing. But rather than serving as an asset, S's phenomenal memory worked to his disadvantage. "Trying to understand a passage, to grasp the information it

We all remember the *Challenger* disaster of 1986—but not as accurately as many of us would like to believe. Where were you and what were you doing on that fateful day? Current recollections can differ markedly from those of only five years ago—or of five years from now. Human memories are not videotapes.

contains (which other people accomplish by singling out what is most important), became a tortuous procedure for S," wrote Luria. Clearly, the perfectly normal process of forgetting must serve as a control on even the most powerful memory.

The 17th-century philosopher and essayist Francis Bacon considered similar exhibitions of memory mere diversion. "I make no more estimation of repeating a great number of names or words upon once hearing than I do of the tricks of tumblers," he wrote, "the one being the same in the mind that the other is in the body, matters of strangeness without worthiness." While Bacon might have gone a bit far in his criticism, he had a point: Memory must be balanced against other operations of the mind.

Normally, short-term memory capacity for relatively meaningless information is limited to about seven random digits and four or five random words or letters. But it can be dramatically increased through practice. In the 1980s, for example, a college student spent one hour a day for three to five days a week improving his ability to remember numbers. After little more than a year, he had increased his digit span to about 80 numbers. He did this by mentally converting the numbers into personally meaningful units, such as the birthdays of friends, historical dates and times, his own track-meet performances, and other personal associations.

Memory experts routinely employ the same method, imposing meaning upon strings of mean-ingless items. In most instances they convert the material to be remembered into a visual image. In fact, that's just what S did.

"When I hear the word *green*," S told Luria, "a green flowerpot appears; with the word *red* I see a man in a red shirt coming toward me; as for *blue*,

this means an image of someone waving a blue flag from a window….Even numbers remind me of images. Take the number *1*. This is a round well-built man; *2* is a high-spirited woman."

Whatever method is chosen, the ability to memorize depends on forming rich and elaborate associations with the items to be remembered. In addition, good memories are based on "study, order and care," in the words of the 16th-century Dutch humanist, Erasmus. Most "memory failures" actually represent failures in attention. For instance, if you arrive at a cocktail party while mentally preoccupied with earlier events at the office, you'll likely quickly forget the names of the first few people you're introduced to. In your distracted state you simply aren't paying sufficient attention to exert the study, order, and care required to form meaningful associations. As a result, you may worry unnecessarily the next morning about your "poor memory."

Also, disruptions in key brain structures can result in serious memory difficulties, because they interrupt the normal process of memory encoding. Obviously, as with the cocktail party experience, you can't retrieve something you don't encode. Specifically where and how does encoding and memory formation take place? Neuroscientists have learned much from the unhappy experiences of two people with brain injuries.

The first, H.M., underwent an operation to stop the spread of epileptic seizures from one cere-bral hemisphere to the other. His surgeon removed the amygdala, uncus, hippocampal gyrus, and the anterior two-thirds of the hippocampus on both sides of the brain. As the result of this operation,

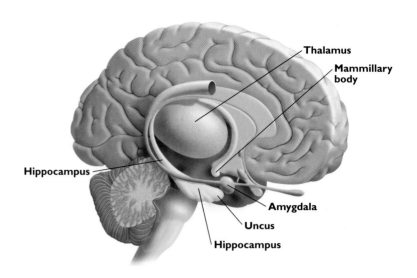

H.M. suffered impairment of his long-term memory. (To neuroscientists, *long-term memory* refers to any memory that is stored by the brain.) A few minutes after a conversation, H.M. couldn't recall any of the details, or even that a conversation had taken place. The other aspects of his thinking—I.Q., reasoning, abstraction, perception—remained unchanged. Only his ability to form new memories was affected.

Another patient, R.B., suffered brain damage during cardiac bypass surgery. As with H.M., his ability to form new declarative memories was impaired. Later examination of R.B.'s brain revealed bilateral damage to the hippocampus. The amygdala and surrounding cortical area were normal, suggesting that hippocampal damage alone can decrease our ability to form new declarative memories. Both H.M. and R.B. retained their procedural memories. In one experiment, H.M. showed improvement in working through a paper-and-pencil maze test, despite his total inability to recall having taken the test from one session to another.

Based on studies of these and other patients, neuroscientists now believe that the hippocampus is involved in the formation and consolidation of declarative memories. The amygdala provides emotional coloring. We don't just remember facts like a computer; we also recall the emotions that accompanied the formation of our memories. On occasion these emotional components can dominate the memory process—as in post-traumatic stress disorder—or they can go missing altogether, as occurs with traumatically induced amnesias, which require hypnosis or other aids for recall.

Alcohol abuse accounts for another long-term memory impairment. Alcoholics often fail to eat an adequate diet, and can develop an acute deficiency

of vitamin B1, or thiamine, a deficiency that harms the brain. This condition, named Korsakoff's syndrome in honor of a 19th-century Russian neuropsychiatrist, is marked by damage principally to the mamillary bodies and the medial thalamus. Like H.M. and R.B., sufferers of Korsakoff's cannot form new declarative memories. In addition, they often make up stories to fill their memory gaps.

Our ability to transform new experiences into long-term memories is highly reliant upon normal functioning of a network of brain structures that include the hippocampus, temporal lobes, and their connections to other midline structures. H.M., R.B., and those who suffer from Korsakoff's syndrome can't form new memories. But they remain capable of remembering events that occurred years before their brains became injured. How is this possible?

———————

Let's say you want to improve your memory. In order to perform at your best you follow Erasmus' advice, proceeding with order, care, and studiousness. You use some associational techniques to transform a string of meaningless words or numbers into a meaningful pattern. At this point you've established the material in your long-term memory and can make statements about the material (declarative memory). How and where is the material now stored in the brain?

To answer the "how," think for a moment about forgetfulness. Ordinarily, you don't forget something suddenly, in its entirety. You forget by a process whimsically referred to as "graceful degradation." For instance, your memory of your college graduation doesn't just drop completely off your memory radar. Rather, your recall of that day loses some sharpness

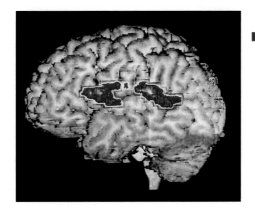

and detail with each passing year. The process isn't terribly different from photographs of the event: subtle losses in clarity, intensity, and definition. We forget slowly, often imperceptibly, because our long-term memories are distributed throughout the brain rather than in specific localized areas, as short-term memory is. We know this because of the pioneering work of researcher Karl Lashley.

In the 1950s, Lashley set out to discover what he termed the "engram" or storehouse of memory. After training rats to run through mazes, he started removing parts of their brains, small sections at a time. He theorized that the rats would suddenly lose their ability to negotiate the maze when he identified and removed the brain area containing the engram for maze learning. To his surprise, Lashley discovered that large-scale removals were required before the rats' maze-running abilities became impaired. And their memory loss was neither sudden nor total. Just like a memory of college graduation, the rats' maze-running proficiency was distributed throughout the whole brain rather than restricted to any one area.

Initially, a rat learns to negotiate a maze using long-term memory, by activating its hippocampus. The maze-running information then becomes widely distributed throughout the rat's brain. Rats with bilateral hippocampal damage never acquire maze-running talents, because they can't establish that initial and crucially important interaction between the hippocampus and other brain regions that are needed to process and store long-term memory. Think of the hippocampus as the brain's entry point for the encoding of information into memory.

In contrast to declarative memory, procedural memories do not seem to require the integrity of the hippocampus. Motor skills such as bicycle-riding reside principally in the premotor cortex and structures beneath the cortex, such as the basal ganglia and cerebellum. After learning to ride a bike you don't have to consciously think about how to pedal, steer, or maintain balance. You just get on and let your brain run a bike-riding "motor program." Since conscious effort and even awareness aren't required for biking and many other activities—except when first learning them—the programs for these activities are best left to the subcortical areas.

————————————

Declarative memory, in contrast, is mainly a cortical process. Some of the fibers from the hippocampus are widely dispersed, while others link to specialized areas. Memories of faces and objects reside within the temporal lobes; landscapes and patterns are relegated to the parietal lobes; and social encounters wind up in the frontal lobes. But since we remember integrated experiences—not just the separate sights, sounds, and other sensations associated with an event—all these memory components must be linked together into a whole. This linkage is accomplished by the association cortices, which occupy the lion's share of the neocortex.

This rich network of association fibers is responsible for the richness and complexity of our memory. Because of it, sensory experiences can

■ FOLLOWING PAGES: Old photographs and faded mementos provide a paper trail for one American's well-lived life. Memorabilia gives us a past—by stimulating specialized memory areas in the brain.

activate memories that have remained silent for decades, beyond our capacity for conscious recall. Writers like Marcel Proust intuitively recognized this mysterious aspect of memory. When, as an adult, he tasted some *petites madeleines*, little cakes that he had used to eat as a child, he experienced a flood of recollections, leading to his pivotal insight into the nature of memory: "It is a labor in vain to attempt to recapture the past. All the efforts of our intellect must prove futile. The past is hidden somewhere outside the realm, beyond the reach of intellect, in some material object."

But while some memories can be recovered by sensory associations, others can be altered by suggestion. Imagine testifying in court about a car accident you witnessed at an intersection controlled by four stop signs. The lawyer asks, "Which car failed to stop for the red light?" If you don't correct the error in his question, you may subsequently remember the accident as involving a traffic light rather than a stop sign. Such malleability in the formation and retrieval of memory accounts in part for the notorious unreliability of eyewitness testimony. Preconceptions, prejudices, unexamined assumptions—all influence memory.

Here are a few other false memories, drawn from psychological experiments:
• Adults were asked about childhood events including one that hadn't happened, such as getting lost in a shopping mall. After being "reminded" by relatives who were in on the experiment, several subjects not only "remembered" the false event but also elaborated on the circumstances.
• College students who at first failed to recall a false event from childhood (this time, an overnight hospital stay for an ear infection) were encouraged to think some more in order to come up with more information before the second interview. One week later, 4 out of 20 students recalled detailed information about their fictitious hospitalization.
• Preschool-age children were told they had undergone a traumatic event: They had caught a finger in a mousetrap and had been taken to the local hospital. Eventually, the children elaborated such highly detailed narratives that some psychologists who specialized in interviewing children were convinced by the narratives that the imaginary mousetrap experience had actually happened!

"Memories for personally experienced events can be altered by new experiences," explains Stuart M. Zola, a psychologist at the University of California at San Diego. "Perfectly detailed memory, flawlessly preserved through decades, sitting unchanged as if in a time capsule, appears unlikely to occur."

Obviously, memory involves far more than simple storage. Our minds don't behave like computers or robots. We not only remember, but also experience emotions when we remember. As the American psychologist and philosopher William James observed, "Memory requires more than a mere dating of a fact in the past. It must be dated in *my past*. I must think that I directly experienced its occurrence. It must have warmth and intimacy."

While James's point certainly holds for personal autobiographical memory—recognizing and recalling the identity of a childhood friend, for instance—does the remembrance of less personal facts involve a similar process? Can you tell me, for example, the identity of the 16th President of the United States?

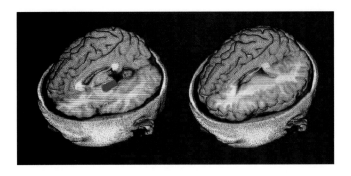

When memory fails: A woman was asked to remember an oral list of words. Ten minutes later, she was given a written list and asked to detect any changes. These PET/MRI scans reveal an active hippocampus (far left), indicating that she is remembering. But her temporal-parietal region, which distinguishes true memory from false memories of words thought to have been spoken, is inactive (left).

If you majored in American history, the answer will doubtlessly occur to you swiftly, without any semblance of James's warmth and intimacy. But if you've devoted little attention to American history, your retrieval of Abraham Lincoln is likely to involve mentally rummaging through memories of personally encountered people, places, and events. Included here might be the remembered details of your high-school history class, your teacher, or your now-yellowed notebooks full of faded handwriting. And, when you come up with an answer, how certain are you of its correctness?

To the former history major—now corporate lawyer—the certainty approaches 100 percent. She would bet her Porsche on it. But the person who spent little time studying history is far less certain, and experiences a vague and visceral uneasiness, a lingering doubt. But then he recalls a specific reference encountered years ago that firmly identifies Abraham Lincoln as the correct answer; suddenly his anxiety is replaced by the "warmth and intimacy" referred to by James.

Part of the difficulty in recalling details from our earliest years is due to normal amnesia. Sit with your mother and page through an album of your early baby pictures. Most likely you won't remember anything pictured or, for that matter, anything that happened before age three or four. Did you repress all those experiences, as has been suggested by some Freud-influenced psychologists? More likely, your amnesia results from the immaturity of your brain at the time of the event. Principally, your cerebral cortex serves as the repository for permanent memory storage. During the first two years of life, these areas are maturing and so remain unavailable for the construction and retention of conscious

memories that can be recalled later. In addition, early language is primitive, incapable of forming a declarative memory.

Another form of amnesia, source amnesia, is extremely common. While most of us can correctly identify Lincoln as the 16th President of the United States, few of us can recall the precise occasion when we learned that fact. That's source amnesia: We remember a fact or an occurrence, but can't remember the source of our knowledge.

Neuroscientists link source amnesia with the frontal lobes. These are the last areas of the cortex to mature, and among the most vulnerable to the ravages of age. As a result, source memory failures are greatest among children and the elderly. In most instances, source amnesia doesn't create great difficulties. But in a courtroom or other formal settings a lot may be riding on a person's ability to recall the exact details of when and where one person said something to another.

Other forms of amnesia aren't normal at all. A particularly intriguing form of amnesia that I've encountered on several occasions in my neuropsychiatric practice is transient global amnesia. As the name implies, this is a brief loss of memory, usually for place and purpose. Typically, the affected person will suddenly ask his companion "Where are we? Why are we here?" Often he will exhibit agitation, as one might expect from someone finding himself adrift in strange and puzzling surroundings. Several hours later, the amnesia disappears as quickly and mysteriously as it arose—much to the relief of patient and companions alike. Although no cause is usually discovered, most neurologists believe such

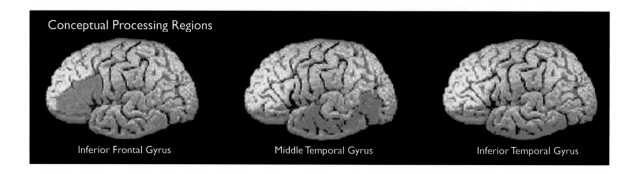

Conceptual Processing Regions

Inferior Frontal Gyrus   Middle Temporal Gyrus   Inferior Temporal Gyrus

attacks result from temporary disturbances in blood flow to the brain—largely because most episodes of transient global amnesia occur in middle-aged or elderly people, who commonly experience such blood-flow imbalances.

Another form of amnesia, now rare, that I've encountered is the so-called fugue state. One of my patients who owned and supervised a large farm about 100 miles outside of Washington, D.C., suddenly "came to" while signing into a hotel in Connecticut. When he expressed puzzlement and distress at how he had traveled the several hundred miles from his home, the desk clerk called the police. They identified my patient from papers in his wallet and called his home. Later the story came out: He was faced with a mountain of bills, a shaky marriage, and uncertainty about his continued ownership of the farm. In response to such pressures, my patient had entered into a fugue state, a psychologically induced amnesia for the events of the preceding days.

As one fugue expert describes the process, "The patient [takes] a two-fold flight from an unbearable life situation—a physical flight so that he is some distance from his appropriate habitat, and a psychological flight into a dissociative state."

In the absence of such afflictions, most of us remember well enough to get through our day. The one glaring exception is the dreaded ravager of memory known as Alzheimer's disease.

Alzheimer's kills about 100,000 Americans a year. It currently affects 4 million, and that number is expected to swell to 14 million by 2040. It is a degenerative disorder of the brain that over 5 to 15 years robs its victims of their ability to work, think,

reason, and—most strikingly—remember. First described in the 19th century by neuropathologist Alois Alzheimer, the illness has probably existed for hundreds if not thousands of years, under a simpler name: senility. A description by the essayist Montaigne is suggestive of the illness:

"Sometimes it is the body that first surrenders to age, sometimes, too, it is the mind; and I have seen enough whose brains were enfeebled before their stomach and legs; and inasmuch as this is a malady hardly susceptible to the sufferer and obscure in its symptoms, it is all the more dangerous."

Usually, Alzheimer's begins with subtle impairments of memory that progress to not-so-subtle ones. Confidence in the correctness of one's memory also undergoes a gradual and progressive erosion. Over the space of only a few years the memory completely fails, even for familiar people, places, and events. Forgetfulness, usually in the form of absentmindedness, gives way to momentary confusion, failures in identification of friends and relatives, compulsive questioning and repetition, and, in the terminal stages, loss of ability to remember even the simplest words. Eventually the Alzheimer's patient is reduced to a shadow of himself or herself—the disease is slightly more common among women—and often succumbs to pneumonia or other complications of a bedridden state.

The mental and physical deterioration of Alzheimer's is accompanied by changes in the brain that include loss of brain substance, degeneration and loss of neurons, accumulation of fibrous deposits within some nerve cells in the cortex of the brain (neurofibrillary tangles), and deposition of other fibrous materials outside of the nerve cells (amyloid or senile plaques). Although much has been learned

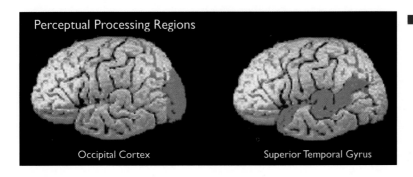

Perceptual Processing Regions

Occipital Cortex   Superior Temporal Gyrus

■ We possess many types of memory, and our brain processes them in different ways. Conceptual memory, for example, emphasizes meaning, while perceptual memory focuses on physical features. These scans show (from left) three regions involved in conceptual memory and two active during perceptual processing.

about Alzheimer's in the last ten years, neuroscientists still aren't certain of its cause. The risk for the disease is increased by a positive family history, earlier significant head trauma, and limited early education. Regarding that last factor, it's speculated that brain health depends upon a "use-it-or-lose-it" formula, with early education providing the foundation and stimulus for lifetime learning.

A person who has a first-degree relative—a biological mother, father, sibling, aunt, or uncle— who is afflicted with Alzheimer's has nearly a 40 percent risk of acquiring the disease by age 90. While this strongly suggests that genetic factors are important, the mode of transmission is usually more complicated than a simple dominant trait. To explore the genetics of this illness, neuroscientists have had to drop down several levels from the behavioral to the anatomic to the chemical and, finally, to the molecular.

———————

Chemically, Alzheimer's patients lack normal amounts of the neurotransmitter acetylcholine, especially in the hippocampus and the nucleus basalis of Meynert, a collection of neurons that give rise to acetylcholine-transmitting cells in the cortex and hippocampus. All current drugs for Alzheimer's work by increasing acetylcholine levels within the brain. Unfortunately, they don't cure the illness, they only slow its progression.

The next level of inquiry, molecular genetics, offers insights that could lead to an eventual cure. So far, four distinct genes—located on chromosomes 21, 19, 14, and 1—are known to predispose a person to Alzheimer's. This knowledge provides researchers with a starting point for their research

into the chemical events leading to nerve cell death. Neuroscientists long suspected that a mutation on chromosome 21 might account for an inherited form of Alzheimer's. An extra copy of chromosome 21 is found in adult Down's syndrome, an illness sharing similar physical features with Alzheimer's: plaques, tangles, and severe memory impairment. In addition, people with Down's syndrome have trouble converting short-term memory to long-term. The prominence of amyloid deposits in both illnesses suggests to researchers that Alzheimer's might result from the mutation of a gene encoding for a chemical precursor of amyloid. That gene occurs on chromosome 21.

Abnormalities in the processing of the amyloid precursor are a common feature in most forms of Alzheimer's. But neuroscientists have a long way to go before completely understanding the illness on a molecular basis. Presence of faulty genes on the four chromosomes, for example, accounts for only half of the known cases. Nevertheless, the discovery of a molecular basis for Alzheimer's holds great promise for future memory research.

Eric Kandel, a neuroscientist at Columbia University, conducted groundbreaking experiments on memory using the marine mollusk *Aplysia californica*. His research concentrated on discovering the

■ FOLLOWING PAGES: Like riding a bicycle or skating on an ice rink, driving a car in traffic involves procedural memory: Once you learn all the controls, you remember how to use them and do so with less and less conscious concentration, in effect rolling along on auto-pilot.

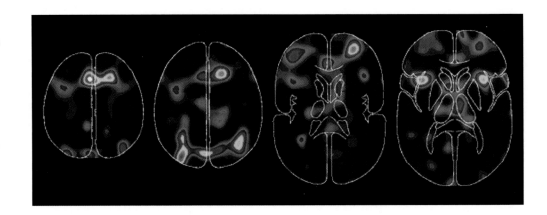

chemical underpinnings of the memory for fear. *Aplysia's* beauty lies in its simplicity. It's a great research subject because it has only about 20,000 neurons in its central nervous system (compared to the hundreds of millions that make up the human brain). Also, its neurons are large and can be easily identified by their size and location.

Like other mollusks, *Aplysia* has a siphon, a sensory structure linked to its respiratory organ, the gill. Touch it, and *Aplysia* briskly withdraws both siphon and gill. But this is not a robotic reflex; since it is a living creature, *Aplysia* can establish what corresponds to memories of its experiences. If its siphon is touched repeatedly, the withdrawal response weakens and eventually stops altogether. It's as if this animal remembers the harmlessness of the earlier siphon touches.

But, Kandel found, if a touch to the siphon is paired with a single strong electric shock to the tail, the creature quickly and powerfully withdraws the gill in preparation for escape. More important, it fails to habituate to this situation. In fact, the more often the shocks occur, the stronger and more lasting the gill withdrawal. If the combination of siphon touch and tail shock is repeated often enough, the withdrawal reflex will be enhanced for periods that range from hours to days or even weeks.

On the molecular level, here's what happens: A single stimulus to the tail activates a chemical signaling system within *Aplysia's* brain, which relies on serotonin. This neurotransmitter strengthens—but only for a few minutes—connections between the sensory and motor cells responsible for gill withdrawal. Repeated stimulation of the tail leads to a pinball cascade of chemical reactions. At the neuron's surface, serotonin acts to increase the level of an intracellular signal known as cAMP. As stimulation continues, cAMP activates an enzyme that enters the nucleus of the cell and attaches itself to proteins that bind to DNA. When this occurs, a gene is triggered, causing other genes to switch on and, in turn, initiate yet another cascade of chemical activity, which leads to the formation of a protein capable of promoting the growth of new synaptic connections. In essence, *Aplysia* reinforces specific neural pathways, changing them through experience.

In humans, short-term memory involves only transient alterations; that's why the information is soon forgotten. But our long-term memories rely on long-lasting chemical and anatomical alterations within the brain. The potential for creating these alterations extends over our lifetime.

Kandel's research, along with the work of others who study the molecular underpinnings of memory, suggests that memory storage within the brain depends upon patterns of electrical and chemical activity among neurons. How do these patterns modify neuronal circuitry and establish memories?

In 1949, Canadian psychologist Donald Hebb suggested that coordinated and repetitive activity between any two neurons strengthened their connections. The same holds true when many neurons are linked into a circuit. Neurons that are active together retain stronger synaptic connections; when one neuron is activated, it tends to rouse the others, which collectively re-create the original pattern. It's been said, "neurons that fire together wire together" and "neurons that play together stay together."

This is the basis of the Hebb postulate, which explains memory as a reactivation of the same pattern of neurons that were activated at the time of the original experience. Each time you think back to your college graduation, the same neuronal pattern is activated. Over the years some of the neurons drop out, either from death or because they've become incorporated into circuits that represent different experiences. This results in forgetfulness for some aspects of the experience and a loss of clarity for other aspects.

Similarly, neuronal networks provide the infra-structure for the formation of habits. Each repeat activation of a particular neuronal circuit facilitates the next activation of the same circuit. Each time a smoker reaches for a cigarette, it becomes easier to reach for the next. Habit modification requires, at the cellular level, a *lessening* of strength of the synaptic connections in the circuitry for that habit.

It's likely that individual neurons are involved in many different circuits and thus participate in many memories simultaneously. Such an arrangement is consistent with the observation that a particular memory can be activated by a host of triggers, such as Proust's *petites madeleines*. The number of specific connections between neurons provides more than enough "hardware" for the establishment of more memories than you could hope to form in a lifetime.

If we accept that the human brain contains something on the order of 100 billion neurons and 100 trillion synapses, and if we assume that memory involves small adjustments to the strengths of these synapses, then the number of possible combinations truly boggles the imagination. Storing information even at the low average rate of one bit per synapse means that the brain would generate 10^{14} bits. (A supercomputer commands a memory of about 10^9 bits.) Now add to this complexity the result of synaptic strengthening brought about through frequent use. The result is like changing from a simple on-off switch to a rheostat, which permits a light to be dimmed or brightened along a rich continuum. Furthermore, the strength of synaptic connections among the neurons in a circuit varies from moment to moment, according to use.

As you learn new things throughout your life, your brain retains sufficient plasticity to encode the memory of each experience within its structure, chemistry, and genetics. Many of these changes can be objectively demonstrated. If a monkey is trained to use a specific finger to carry out a specific task and that task is repeated thousands of times, a larger region of the cortex will contain neurons involved with that finger. The other, less frequently used fingers will, in turn, have smaller cortical regions dedicated to them.

A similar enhancement of brain activity accompanies the learning and memory of skills in humans. As noted earlier, areas of the motor cortex that control hand and finger movements are much bigger in professional keyboard and string players than they are in non-musicians. What's more, the increase in size within the motor cortex relates to the age at which the musician began training. The earlier the training, the bigger the musically relevant areas of the motor cortex.

Memory also is affected by aging, as anyone over 40 can no doubt attest. Typically, we have difficulty with the rapid retrieval of specific words. Information is still retained; it just takes longer to come up with it. Psychologists refer to this fall-off in memory performance as age-associated memory impairment, or AAMI. It's perfectly normal and is not a herald of Alzheimer's disease.

This lessening of memory speed is part of the general slowdown in responsiveness that comes with age. We remember more slowly just as we move more slowly. But before getting depressed about this inevitable aging response, remember the trade-offs.

Despite protestations to the contrary (often made while looking in a mirror), most people over 40 would have reservations about suddenly being a 20-year-old again. For one thing, regaining your youthful appearance would mean losing all the comforting possessions you've accumulated over the years—along with a return to the bottom of the occupational ladder. Gone, too, would be all your accomplishments as an adult. Most painful would be the loss of all the memories accumulated over your lifetime, those links with the past that form the basis of your very identity. You'd also have to surrender all the knowledge that you've gained from a lifetime of observing people and events, the *wisdom* you've acquired.

Dr. Paul Baltes, who studies aging, defines wisdom as "expert knowledge about life in general and good judgment and advice about how to conduct oneself in the face of complex, uncertain circumstances." Of course, wisdom also includes rich factual knowledge, the ability to make judgments while taking into account contexts, values, goals, and priorities—and a recognition that everything is temporary, relative, and unpredictable. As a Hollywood icon might put it, "Don't sweat the small stuff—and in the Big Picture, it's all small stuff."

Baltes's research, carried out with his late wife Margret, supports the notion that wisdom is a domain where the old can excel. "Older adults seem to have acquired the dispositions and skills to benefit from social exchanges with others to solve the dilemmas of life. Here may lie the foundation for the many success stories of grandfathers, grandmothers, and older mentors who are able to express warmth, understanding, and guidance."

Your brain is dynamic, changing from moment to moment as a result of your experiences. Although embedded within the architecture of your brain, your memories remain modifiable by subsequent experiences, inclinations, even moods. When depressed, we tend to dwell on sad or otherwise negative aspects. As depression lifts, memories of the good times seem to return as if by magic. That's why it's important to maintain a positive outlook and, if depressed, seek help. Prozac and other, newer drugs that modify serotonin and other neurotransmitters induce a more positive mood while at the same time enhancing memory.

Certainly, everything possible should be done to conserve and enrich our memory abilities. We can strengthen them simply by employing reverie or other relaxed efforts aimed at mentally re-creating life's joyful moments. We can also practice memory-strengthening exercises such as association. But whatever methods we use, we should do all we can to preserve our links with the past. Put at its plainest, *we are our memories.*

■ Clinging to an image of her missing son, an Afghan mother waits as a guard tries to arrange a meeting with officials. The photograph she holds is less fragile but also less complete than the lifetime of memories stored in her brain.

MECHANICS
OF EMOTION

I magine yourself sorting through your mail upon arriving home from work. Nothing special, just the usual mix of bills, requests for money, catalogs, magazines, and...oh yes, a letter from the Internal Revenue Service.

Depending on whether you prefer to hear the good news or the bad news first, you may open immediately or put off until later the reading of that official and potentially ominous communication from the tax people. But one thing is certain: your brain is no longer functioning in a just-casually-sorting-through-the-mail mode. Measurements would reveal that your breathing has become a bit more rapid and shallow, your heart rate and blood pressure are slightly elevated, and certain stress-related hormones are pouring into your blood-stream. Taken together, these responses constitute an anxiety reaction.

Attempts to understand anxiety and other emotions on the biological level began with some early research with cats. When a cat's two cerebral hemispheres are removed, the animal appears angry: the hairs on its back and tail stiffen, its pupils dilate, its blood pressure and heart rate begin to race. The animal arches its back, snarls and bares its claws in anticipation of attack.

Yet all these feline displays of anger disappear if a cut is made severing the cat's hypothalamus from the midbrain. On the basis of just such experiments, physiologist Philip Bard suggested in 1928 that emotional behavior—at least the behavior associated with rage—did not require a cerebral cortex. Subsequent experiments by others further established that the basic circuits responsible for emotional behavior exist in the hypothalamus and other brain areas far below the cerebral cortex.

■ Every recruit's worst nightmare: Testing will and emotional fortitude, drill instructors at Pensacola Naval Air Station verbally annihilate the most recent arrivals to Officer's Candidate School—later rebuilding them in their own image.

We now know that impulses starting in the hypothalamus descend to the brain stem and into the reticular formation, composed of over 100 separately identifiable cell groups. Through its connections with the brain stem, the hypothalamus influences heart rate, breathing, and other organ responses in just about every part of the body. But what arouses the hypothalamus into action? How does reading that notice from the IRS set off responses that you experience as a sense of chest constriction, stomach churning, or an I-can't-get-my-breath form of anxiety?

All emotional experience results from the activation of the limbic system, so named because it forms a ring ("limbic" means border) around the corpus callosum and medial cortex. The limbic system's main components are the hippocampus, the cingulate gyrus, and the amygdala. Two-way communication between the limbic system and the cerebral cortex explains your anxious response to the IRS notice. In fact, neuroscientists now believe it's likely that your anxious response may have begun milliseconds prior to your conscious recognition that you had mail from the IRS.

Joseph LeDoux, of the Center for Neural Science at New York University, writes, "It is possible for emotions to be triggered in us without the cortex knowing exactly what is going on." His research has shown that the amygdala specializes in such preconscious fear recognition. "The amygdala is programmed to react without benefit of input from the thinking part of the brain, the cortex."

LeDoux's rather surprising conclusion is based on many experiments he has carried out on fear conditioning. In a typical LeDoux experiment, a rat isolated in a cage hears a tone just moments before an electric shock is delivered to its feet. Soon the creature starts to respond to the tone alone.

On first hearing the tone, the rat's blood pressure rises and the animal freezes in place, as if too terrified to move. Left at this, LeDoux's experiment would be no more than a rodent version of Pavlov's famous conditioning-response experiments on dogs. But LeDoux's goal was more ambitious: to identify what parts of the brain are absolutely indispensable to the conditioned-fear response.

He was surprised to find that the conditioning response could be established even when the rat's auditory cortex was destroyed: the rat still froze in fear and its blood pressure skyrocketed. But if destruction occurs below the cortex, in the auditory thalamus, the rat cannot be fear-conditioned.

In his book *The Emotional Brain*, LeDoux writes: "The fact that emotional learning can be mediated by pathways that bypass the neocortex is intriguing, for it suggests that emotional responses can occur without the involvement of the higher processing systems believed to be involved in thinking, reasoning, and consciousness."

LeDoux's research suggests there are parallel pathways to the amygdala, "low" and "high" roads that relay information from the outside world. The low one leads directly from the thalamus; the high is longer and less direct, leading from the periphery to the thalamus, to the cortex, and finally to the amygdala. Each route provides different levels of information; the direct one—what LeDoux calls "quick and dirty"—provides rather crude representations. The cortical pathway, in contrast, allows for elaboration and meaning.

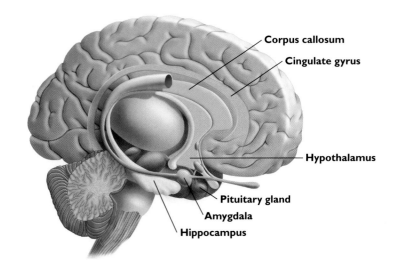

■ Core processing plant for the human emotions of fear, rage, and pleasure, the brain's limbic system focuses on self-preservation, reproduction, and the formation of memories.

Corpus callosum
Cingulate gyrus
Hypothalamus
Pituitary gland
Amygdala
Hippocampus

The proper functioning of this parallel arrangement of impulses to the amygdala can literally mean the difference between life and death. It takes only about 12 milliseconds for a sound to reach the amygdala through the thalamus, and nearly twice as long to take the longer path through the cortex. According to LeDoux, the thalamic pathway "cannot tell the amygdala exactly what is there, but can provide a fast signal that warns that something dangerous may be there."

Thanks to fast processing by the thalamic pathway, we react to the sound of a gunshot by diving for cover—even before we know what's happening. Only later does our sensory cortex provide a plausible explanation for what happened. In the meantime, escape is our number one priority. There's a lot of wisdom in this arrangement. If the sound we hear really is gunfire, our quick amygdala-mediated response may save our life. If the sound is just a backfiring car, our only loss will be the cleaning bill for our suit. As LeDoux puts it: "From the point of view of survival, it is better to respond to potentially dangerous events as if they were in fact the real thing than to fail to respond."

Of course, we aren't always aware of the brief intervals that separate the arrival of information along the two pathways, so we often attribute our anxiety to the moment we consciously recognized the threat. But, as LeDoux's research suggests, the anxiety actually started slightly earlier. Thus your conscious recognition of the IRS letter (a cortical response conveyed along the high road) lagged slightly behind the anxiety generated by your having vaguely, unconsciously detected it (by way of the direct low road, from thalamus to amygdala). That's why you felt vaguely and inexplicably anxious while

sorting through the mail. In fact, your anxiety began milliseconds before you consciously recognized the source of the IRS letter.

All of us have experienced a vague anxiety prior to discovering the cause of that anxiety. Perhaps you find yourself suddenly and inexplicably tense at a cocktail party. A moment later you understand your anxiety as you become aware of glares from across the room—coming from your former wife and her new husband. Your amygdala appraised and responded to the situation an instant before your cortex could provide the explanation.

Anxiety disorders can result from disturbances along the low road. If fear becomes associated with a specific experience like going shopping, this phobia may be difficult to reverse through reasoning or discussion. This may be one reason why psychotherapy alone is rarely successful in the treatment of phobias or any of the other five major anxiety disorders.

A particularly severe form of anxiety disorder known as post-traumatic stress disorder (PTSD) involves reliving the vivid and emotionally wrenching memories of some horrifying experience from the past. As a typical symptom, the PTSD sufferer may startle in response to a loud noise.

To most of us, a slamming door is annoying and, if loud or close enough, mildly startling. Unless we've had some experience with weapons, we don't usually mistake it for the firing of a gun. But to a Vietnam veteran who saw dozens of his comrades killed or wounded, that slamming door may activate a full-fledged stress response, causing him to dive for cover or break out in a sweat. His blood pressure and pulse rates rise; his hypothalamus goes into high

■ Letting loose all emotion, a Lakota family on South Dakota's Pine Ridge Reservation grieves for two of its members—one a newborn—who died in a car accident. Often more difficult to bear than physical pain, emotional trauma can ravage the mind over time.

**Emotion is always in
the loop of reason.**
—Antonio Damasio

gear, nudging into action organs of the endocrine system, which release stress-related hormones into the bloodstream. But the most painful part of PTSD is the accompanying sense of vulnerability and helplessness. The symptoms can't be willed or wished away; no preventive measures exist that can prevent future occurrences.

LeDoux suggests that trauma somehow "biases the brain in such a way that the thalamic pathways to the amygdala predominate over the cortical ones, allowing these low-level processing networks to take the lead in the learning and storage of information."

Panic attacks—feelings of nameless dread and fear of imminent death that come out of the blue— are similar to PTSD but aren't necessarily conditioned by previous experiences. Internal rather than external events act as the triggers. PET scans taken during panic attacks reveal changes in blood flow in the temporal lobe, hippocampus, and amygdala. It's thought that panic attacks involve activation of the amygdala and its projections to the sympathetic nervous system, responsible for the body's fight-or-flight response. We sense that something is wrong, but aren't sure what. Uncertainty and anxiety build and then explode in a full-blown panic attack.

———————

Not everyone becomes anxious when exposed to the same stresses and uncertainties of modern life. Genetic and temperamental variables undoubtedly play a role in anxiety and fear conditioning. Infants only a few months old show striking temperamental differences; some respond to new situations with agitated movements of their arms and legs, archings of the back, and fretting and crying outbursts. Others respond hardly at all.

After studying anxiety in children for years, Harvard psychologist Jerome Kagan noted, "The children who showed high fear at all ages were most likely to develop anxious symptoms by the time they were seven."

What's more, the tendency to develop anxiety appears to run in families; anxious parents tend to produce anxious children. At this point, neuroscientists aren't certain whether genetics or exposure to anxious family members is the major contributor. Either way, once an anxious temperament becomes established, it leads to anxious preoccupations even in situations lacking a reasonable basis for anxiety.

Northwestern University psychology professor and researcher Susan Mineka writes, "In people who are anxious their attention is drawn toward threatening stimuli in their environment. They tend to overestimate the likelihood of bad things happening to them and also tend to interpret ambiguous information in a negative manner."

Post-traumatic stress disorder exemplifies Susan Mineka's point. A person with PTSD retains low-level feelings of anxiety that rapidly escalate into a full-fledged attack when he encounters something threatening in his environment. A full explanation requires correlating events at the social, psychological, chemical, and anatomical levels.

Naturally, we experience emotions other than fear and anxiety. We can be happy or sad, pleasantly contented or filled with anger. Are different brain areas dedicated to these different emotions? Does the variety in emotional experience and expression from one person to another derive from differences in the structure or operation of one or more parts of the brain? Neuroscientists answer both questions with a qualified yes. They believe that, rather than

One of humankind's oldest emotional reactions, fear of snakes puts the limbic system of many of us into overdrive. But some, like this south Florida reptile poacher, overcome such innate fears.

having a single center for all emotions, the brain is set up so that some areas handle specific emotions.

In general, the left hemisphere is concerned more with positive emotions, the right hemisphere with negative emotions. Damage to the brain's left side often results in symptoms of depression such as crying, agitation, and expressions of hopelessness. Damage to the right side of the brain, in contrast, often induces an inappropriate cheerfulness and a tendency to make light of the damaging effects of the injury or even deny them altogether.

———

Hemispheric differences in mood also occur in people with perfectly normal brains. If a person listens through earphones to recordings of human speech, the right hemisphere is better than the left at detecting subtle emotional nuances, such as mood-associated variations in speech. Similarly, if pictures of human faces are selectively flashed to either the right or left field of vision, the emotions depicted are more quickly and accurately identified from what's seen in the left hemifield (that is, the portion of the visual field processed by the right hemisphere). At least for right-handers, emotions are primarily expressed through movements of the left side of the face—which is controlled by the right hemisphere. Thus the right hemisphere is, in general, more directly associated with the expression and perception of emotions than is the left.

Research—by Richard J. Davidson, professor of psychology and psychiatry at the University of Wisconsin-Madison, and others—provides one explanation for this mood-associated hemispheric asymmetry. In an experiment using normal volunteers, Davidson showed pictures of what were

deemed either positive images (a child playing in her yard, for example) or negative images (such as a bloody scene of a car accident). Repeated display of negative images induced marked changes in the mood of the participants; they reported feeling "sad" or "low," and their facial expressions veered toward sadness. As the mood worsened, an accompanying change occurred in the brain's regional glucose metabolism, as assessed by PET scanning.

The scans appeared similar to those of people who had suffered damage to their left hemispheres. With that hemisphere less active, the right one took control. Additional measurements of cerebral blood flow showed activation of the amygdalas on both sides of the brain, a finding that, to Davidson, "confirmed the amygdala's involvement in negative stressful emotions."

In another experiment, Davidson used people who showed a tendency for either extreme left-sided or extreme right-sided activation in their prefrontal cortices. He asked each person to select from a list of positive or negative adjectives those words that best described their feelings about themselves. Davidson discovered a correlation between the words chosen and a person's prefrontal activation pattern. The subjects with left-sided activation felt more positive about themselves, as measured by the adjectives they selected, such as "confident," "happy," or "excited." What's more, left prefrontal activation appeared to be associated with other indicators of

FOLLOWING PAGES: Another powerful and long-lived human emotion—love—works its magic on a Russian couple as they unwind aboard the Trans-Siberia railroad.

■ Neurotransmitters of emotion: Norepinephrine (right) and dopamine (far right) play key roles in the propagation of various moods and behaviors.

reduced stress and anxiety. So the left prefrontal cortex is involved in positive emotions; the right deals with negative ones.

Each side coexists in a balance that varies from person to person. In people who have basically negative, anxious, and fearful approaches to life, the right hemisphere is dominant. Positive, confident people show left-sided predominance. An individual's pattern of right-left hemisphere activation involves any number of factors, including genetics, experience, and brain organization.

At the Library of Congress in 1998, Davidson summarized his findings: "Our imaging work suggests that the prefrontal cortex—particularly the left prefrontal cortex—may be important...in dampening the response to negative events, and particularly in shutting off the negative response quickly once it has been activated." Davidson believes that the left prefrontal areas accomplish this by inhibiting the amygdala.

His experiments also provide a neurological explanation as to why brain-damaged patients respond so differently, depending on the side of the brain that's injured. Whenever one side is damaged, the other side of the brain operates unopposed. Thus damage to the left prefrontal cortex—which generates positive emotions—leaves a person vulnerable to the effects of the right prefrontal area that, now unopposed, expresses anxiety and other distressful negative emotions.

Davidson and other researchers suspect that people with enhanced left-sided activation are better able to turn off negative emotions and accentuate positive ones. They are less likely to become depressed and, if they do suffer depression, they usually recover more quickly. It's also interesting to

note that antidepressant drugs increase activation of the left prefrontal cortex. But even people who don't take such medications can accomplish this shift in balance through therapy sessions and other psychological treatments.

Davidson's findings provide intriguing confirmation of our grandmother's advice to "Try to think good thoughts and stop your worrying." But even this simple recipe can be taken to harmful extremes. Imagine what it would be like to experience no anxiety at all; no matter what happens, you feel no inner discomfort. A great state of mind to be in, right? Actually, people who never experience anxiety aren't at all lucky, for they must function without the usual early warning signals of danger.

———————————

Consider your likely response to the initial anxiety caused by that IRS letter. After the first anxious moments you would undoubtedly calm yourself with reasoned self-assurances that you've kept careful records, done nothing wrong, and therefore have nothing to fear from an audit. This ability to talk yourself out of the initial anxiety depends upon a smoothly functioning cerebral cortex, especially the prefrontal lobes. Brain-imaging studies suggest that the prefrontal cortices, especially the left, inhibit activity in the emotion-mediating amygdala.

Balance between the prefrontal cortex and the amygdala is an important aspect of mental health. Patients with damage to either structure—or to the pathways connecting them—fail to experience normal emotions. If the damage concerns their prefrontal lobes, they fail to become anxious under circumstances when anxiety would be appropriate. For instance, we all become anxious at the prospect of experiencing pain. But if our prefrontal cortices are no longer working normally, pain may occur without any accompanying anxiety and suffering.

Neurologist Antonio Damasio first observed this type of absence of anxiety in a patient who underwent an operation that involved cutting the prefrontal fibers in order to relieve severe facial pain. Prior to the operation the man "crouched in profound suffering, almost immobile, and afraid of triggering further pain in his face." Two days later, he was a different person: The doctors found him absorbed in a game of cards and asked his condition. "Oh, the pains are the same, but I feel fine now, thank you," he responded. Concluded Damasio, "Clearly, what the operation seemed to have done, then, was to abolish the emotional reaction that is part of what we call pain."

Damage to certain regions of the frontal lobes also impairs that person's ability to appreciate future negative consequences. People afflicted with such damage often speak and act without tact or consideration for others. If you were frontally damaged, for instance, you might do or say things that on occasion you probably fantasize about but sensibly refrain from carrying out. For instance, if you became annoyed at your boss, you might just "tell him what he can do with this job"—and suddenly find yourself unemployed. If your spouse becomes upset over the death of her mother, you might impatiently advise her to "move on with your life." In fact, two of my patients expressed these sentiments after injury to their frontal lobes. Moreover, they couldn't understand why the boss and spouse reacted so negatively to their outspokenness; their frontal damage had deprived them of the ability to appreciate the effect their words could have.

Anxiety about future consequences prevents many of us from reckless behavior, impulsive decisions, and perhaps even the commission of minor crimes (such as hedging on tax returns). We anxiously imagine ourselves in prison or otherwise deprived of our freedom, reputation, or possessions.

A person lacking anxious feelings doesn't undergo this perfectly normal process. Nor does the process stop there. Once in prison, criminals with frontal-lobe impairments fail to learn from the negative experience of forced confinement. And, lacking normal and perfectly appropriate anxiety, they aren't all that uncomfortable. Thus after release from prison, the prospect of future incarceration does not induce the necessary anxiety to avoid committing additional crimes. Hardened criminals often describe a lack of anxiety about the dangers implicit in a criminal career. Not all suffer frontal lobe damage. But whatever the cause, their failure to experience basic human emotions is what gets them into future trouble.

Damasio characterizes people afflicted with such frontal injuries as "lacking emotional reactivity at a high level." Since they do not experience emotions such as fear or anxiety, he says, they lose "their ability to feel emotion relative to the future consequences of their actions."

All emotions—good and bad, the pleasant and the unpleasant—are mediated by interactions within a circuit that involves structures of the limbic system. The amygdala and the prefrontal cortex both play prominent roles within this circuit. Anxiety serves a useful purpose as a signaling system that calls forth responses aimed at correcting something in our lives. But when this system malfunctions, the result may be persistent and overpowering panic or other anxiety disorders. As we will discuss later, emotional disturbances can be treated and sometimes cured with medications designed to modify the chemistry of brain areas that are involved in the generation and expression of emotion.

Finally, since emotions involve both the frontal cortex and the amygdala (as well as other limbic components), our own personal goals should be to achieve an integrating balance between our thoughts and our emotions. In the words of Joseph LeDoux, "If these nerve pathways strike a balance, it is possible that the struggle between thought and emotion may ultimately be resolved not by the dominance of emotional centers by cortical cognitions, but by a more harmonious integration of reason and passion. With increased connectivity between the cortex and the amygdala, cognition and emotion might begin to work together rather than separately."

Indeed, no emotions are "pure;" they cannot be totally separated from the thinking part of the brain. Thinking and feeling coexist within an interweaving contextual pattern; one cannot be understood without reference to the other. As Antonio Damasio puts it, "emotion is always in the loop of reason."

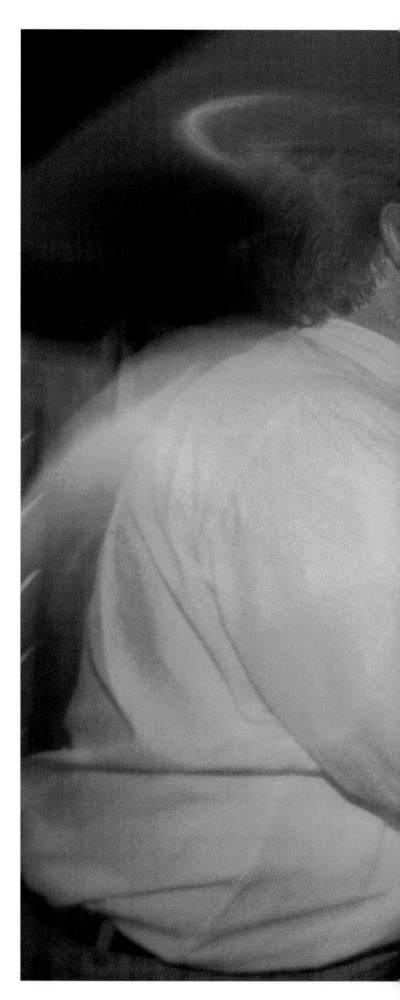

■ Fighting pain with prayer, Pentecostal preacher Wayne Pugh lays hands on the faithful in Ballston Spa, New York. Many participants claim they experience a sense of energy, warmth, or even electricity at Pugh's touch. While science struggles to explain such phenomena, religion has been shown to produce peace of mind and, in some instances, quantifiable health benefits.

Time—and our attempts to control it—can provoke a host of stress-related illnesses, as these Manhattan-bound rail commuters in Darien, Connecticut, can likely attest.

MIND AND BODY TOGETHER

As you sit reading this book, you are mercifully spared any conscious awareness of many critical processes going on in your body. Your heart and breathing rates, the amount of saliva secreted by your parotid glands, the passage of urine from the kidneys into your bladder—these are only some of the body's automatic processes that occur outside of your awareness. We know about how these processes are controlled thanks to the work of a 19th-century playwright-turned-physiologist.

Claude Bernard began his career writing plays for performance in a regional theatre in France. Encouraged by local success, he forwarded one of his plays to a Paris drama critic—who suggested that Bernard forget about the theatre and apply his talents to medicine. Bernard took the advice and in 1858 went on to author a revolutionary concept about the organization of the human body.

He coined a phrase, *la fixité du milieu intérieur*, which refers to his notion that the body maintains a relatively unchanging internal environment regardless of changes outside it. When the temperature drops below freezing, for example, your body mobilizes adjustive mechanisms such as shivering and constricting the diameter of blood vessels in exposed areas; the result is the preservation of body heat. Similarly, if the temperature rises over 100°F on a summer afternoon, your body compensates by perspiring and by dilating your blood vessels—two adjustments that cool the body.

Some 70 years after Bernard, American physiologist Walter Cannon restated the idea, calling it homeostasis and adding that the body's self-control is exerted by the activity of what is known as the involuntary, or autonomic, nervous system.

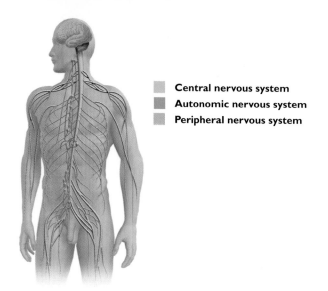

Central nervous system
Autonomic nervous system
Peripheral nervous system

As you may recall from elementary biology, the autonomic nervous system (ANS) consists of two divisions: the sympathetic and the parasympathetic. The sympathetic organizes the body's responses to events that require maximal exertion: the fight-or-flight response. The parasympathetic division, in contrast, oversees the "relax and smell the flowers" approach to things, which physiologists call the relaxation response. Both divisions connect the central nervous system (that is, the brain and spinal cord) with the smooth muscles of the internal organs, blood vessels, and skin.

Fibers of the sympathetic system arise from neurons in the thoracic and upper lumbar regions of the spinal cord. Parasympathetic fibers arise either higher up or lower down, specifically from centers in the brain stem and lower spinal cord. Think of the sympathetic and parasympathetic systems as occupying the two scales of a balance. Too much weight on one, and the balance tips over. If fight-or-flight plays too constant or controlling a role, you live on a knife-edge of anxiety. Too relaxed a response, and you may not take the necessary steps to protect and promote your own best interests.

Certain autonomic responses prevent the appearance of others. You cannot be simultaneously anxious and relaxed or angry and tranquil. That's partly because opposing emotional experiences call into play opposite physiological mechanisms that involve the same organs. When you're angry, the resulting sympathetic stimulation causes your pupils to dilate (so your eyes can take in more information). But when you're in a mellower mood, your pupils constrict thanks to the overriding parasympathetic influences. Similarly, the dry mouth you experience when you're anxious results from the sympathetic

influence on your salivary glands. But with relaxation, the parasympathetic system takes over and increased salivation results.

In order to put all this into perspective, let's briefly review some basic terminology. The central nervous system (CNS) extends all the way down from the very top of the cerebrum to the tail end of the spinal cord. Thus it consists solely of the brain and spinal cord and is totally contained within the bony casings of the skull and spine. Everything other than the brain and spinal cord is referred to as the peripheral nervous system (PNS). Included are the sensory neurons and their receptors, the motor neurons and their extensions, and most of the autonomic nervous system—that is, the sympathetic and parasympathetic systems.

While dividing things into neat packages can help us understand how these parts work, we must remember that the brain functions in an integrative manner, with vast and rapid communications linking one part to another. One of the most important and recently discovered integrations involves the central nervous and the endocrine systems. A system of ductless glands, the endocrine system is functionally organized around the pituitary, the so-called master gland that regulates all endocrine secretions.

Anatomically linked to the hypothalamus, the pituitary and its functions remained mysterious for many centuries. Since it sits just above the nasal sinuses, early anatomists thought it must have something to do with nasal secretion. "Pituitary" is a variation of the Latin for phlegm, and in naming this gland, celebrated anatomist Andreas Vesalius assumed that it discharged mucous.

Structurally, the pituitary is connected to the hypothalamus by blood and nerve pathways. A message from the cortex or limbic system first activates the hypothalamus, resulting in the secretion of chemicals that in turn activate the pituitary. For example, hypothalamic activation can lead to increased production of oxytocin by the pituitary, which causes contraction of smooth muscle in the uterus and breast, and to production of vasopressin, which acts on the kidneys to increase water retention and raise blood pressure. Four different hormones derived from one part of the pituitary control general body growth, the activity of the adrenal glands atop the kidneys, the activity of the thyroid gland in the neck, and the activity of the sex organs.

The hypothalamus is best thought of as a coordinating center that integrates information involving the brain and the endocrine system, which has organs all over the body. It also is involved with the behavioral expressions of emotions. As we saw in the last chapter, direct stimulation of the hypothalamus of a cat elicits an angry response: heightened blood pressure, constricted pupils, arching of the back, and raising of the tail. These responses result from hypothalamic and pituitary activation of the sympathetic portion of the autonomic nervous system. The ANS, in turn, activates endocrine glands such as the adrenals.

The mutual influences of the hypothalamus, the pituitary, and the endocrine glands on each other become especially important during times of stress. When you're late for a bus, for example, the stress you experience sets off a whole series of responses that begins in the cerebral cortex. It goes like this:

The frontal lobes decide on the importance of catching the bus and the consequences involved with missing it, such as arriving late for work. The frontal lobes then issue a "Go for it!" order that activates not only the muscles of the legs but also portions of the limbic system. A message goes down the chain of command from the limbic system to the hypothalamus and then to the pituitary, which releases into the bloodstream a hormone called adrenocorticotropic hormone, or ACTH. Upon arriving at the adrenals, ACTH stimulates these glands to release the stress hormones—epinephrine and corticosteroids, the adrenal cortical hormones. Those hormones, in turn, help you to run faster and just make it onto the bus as it begins to pull out.

At this point, while sitting on the bus, you're aware of the rapid beating of your heart and a breathing rate that's about three times normal. You take a deep breath and settle back. By now, your "take action" response is beginning to wind down. The increased levels of norepinephrine and corticosteroids in your bloodstream are exerting what's called a negative feedback effect on the hypothalamus, inhibiting it from ordering the release of any additional ACTH.

The regulatory feedback system you've just experienced while pursuing the bus is an example of Bernard's *la fixité du milieu intérieur* and Cannon's homeostasis. Working together, your nervous and endocrine systems tend toward an equilibrium. In fact, if anything interferes with that equilibrium, your general health can be in great danger. For instance, if your body responses don't adjust and you feel for the rest of the day just like you did while running for that bus, you would be suffering from a stress response.

No matter how fortunate your life circumstances, no one is exempt from stress. And severe or unremitting stress can be devastating to physical and mental health. Included in the harm resulting from sustained stress are high blood pressure, metabolic disorders such as diabetes, indigestion, and impotency, and an increased incidence of infections like colds and flus.

But the most worrying effect of prolonged and inappropriate stress is the harm inflicted on the brain. Glucocorticoids (released by the adrenals during stress) damage the hippocampus and thus interfere with the initial coding of new information necessary for learning and memory. They also exert damaging effects on neurons elsewhere in the brain and thereby increase a person's susceptibility to strokes, seizures, and infections.

Stress-induced hippocampal damage can be observed in post-traumatic stress disorder, major depressions, and, in some cases, in normal aging. Some neuroscientists even speculate that the increased incidence of Alzheimer's disease may result from the effects of prolonged stress on the brain. Certainly the target is the same: Hippocampal damage occurs both in prolonged stress and in the earliest stages of Alzheimer's disease.

In order to understand and manage stress it's necessary to study fear, since stress is invariably associated with it. And since the amygdala encodes our memories for fear, everything we can learn about this part of the brain and its widespread connections is relevant for our understanding of stress.

Large portions of the brain are devoted to the fear and stress responses, simply because survival is at stake. If you fearlessly return to the location of an attempted mugging that you experienced a few months earlier, for example, you might not have the opportunity for a third visit.

While the amygdala signals danger and stimulates that part of the brain that increases cortisol production for fight or flight, the prefrontal cortex inhibits that same brain area. Both exert action on the hypothalmus, which ultimately controls the secretion of stress hormones, and both enhance our chances for survival by nicely balancing each other—at least in theory. Unfortunately, the balance tends to favor the amygdala. As researcher Joseph LeDoux puts it, "An emotional reaction like fear can more easily gain control over the cortex and influence cortical processes than the cortex can gain control over the amygdala."

But despite our "hard-wired" propensity to favor the fear and stress responses, we're certainly not helpless. It's refreshing to consider that many—indeed most—of the stresses in our lives are self-created and can be eliminated by changes in our attitudes and behaviors. Reducing stress therefore involves developing stress-reducing skills. For instance, several surveys indicate that Americans rate public speaking as the most stressful activity they can be called upon to perform. But, as I have personally discovered, public speaking can be truly enjoyable after additional experience and practice. What's needed is a change in attitude that converts a stress into a challenge.

Overall, our response to stressful situations depends on several factors, among them genetic inheritance, overall physical and mental health, and how we perceive stress. While acute stresses aren't necessarily bad—and can even be beneficial in the

In order to understand and manage
stress it's necessary to study fear, since
stress is invariably associated with it.

—Richard Restak

short term by perking up the body's immune responses—chronic prolonged stress is always harmful. If it builds up over weeks or months, permanent hippocampal damage can occur. Those who suffer from post-traumatic stress disorder often experience smaller than normal hippocampi—and this is only the most easily observed indicator of stress-associated brain damage.

Can we halt or at least reduce our chances of acquiring a stress-related illness? Certainly. Various approaches offer that promise, some of them beginning in childhood. We know that children who suffer abuse or have a parent who is alcoholic, a drug addict, or a victim of spousal abuse are two to four times more likely to come down with serious illnesses like heart, brain, or lung disease later in life. The more stress endured by a child, the more likely he or she will suffer stress-associated illness as an adult. Clearly, whatever can be done to identify and reverse severe childhood stresses will lessen the chances for later stress-associated illnesses.

Stressful experiences encountered later in life take a similar toll. War veterans with post-traumatic stress disorder are more susceptible to a variety of serious illnesses, including alcohol and drug abuse, along with diseases of the lungs, heart, and digestive organs. And once a person becomes sick, stress makes that illness worse. For example, multiple sclerosis (MS), a disease of the brain and spinal cord that predominantly affects socially and occupationally active people between the ages of 20 and 40, often progresses more rapidly among those who are enduring stressful situations. For such people, stress reduction can be every bit as beneficial as presently available treatments for MS. (That is, drugs that decrease the number and severity of acute attacks; there is no cure for multiple sclerosis.)

Such benefits from stress reduction shouldn't come as a surprise. Medical science has recognized for years that our thoughts and emotions can influence our general health, for better or for worse. That's because the brain doesn't exist in isolation from the body but, as noted earlier, maintains active lifelong connections with the endocrine system. What's more, researchers over the past decade and a half have also discovered direct physical links between the brain and the immune system.

Think of your immune system as an army composed of various field units located in special organs widely dispersed throughout your body. After being created within the bone marrow, immune cells move on to the thymus, a gland beneath the breastbone, where they undergo basic training in recognizing and destroying antigens: foreign agents such as bacteria, viruses, and other intruders. Once trained, your immune-cell soldiers take up residence in barracks such as the lymph nodes, the spleen, and other tissues, including the skin, lungs, gut, and liver. And like real soldiers, these new recruits don't just stay put but move around to trouble spots all over your body.

Your immune system and brain work together as a team to retain stable, healthy conditions within the body. The brain, of course, is the master controller, influencing the immune system through two major pathways. Extending from the spinal cord, nerves of the PNS convey communications that originate in the brain to immune organs such as the lymph nodes and spleen. But even more important than these direct connections are the indirect, long-distance, chemical connections that unite the brain and the immune system. Many of the same neurotransmitters found in the brain occur at sites of the immune system; many immune-system chemicals regularly enter the brain. Hormones secreted by components of your immune system affect not only the cells in your endocrine system but also those in your brain. In each instance, chemicals are the common mechanism of exchange.

For example, white blood cells produce proteins called cytokines, which help coordinate different parts of the immune system to fight invaders. The cytokines interleukin-1 and interleukin-2 act as intermediaries, connecting immune cells and other bodily cells, including neurons. When a person is sick, cytokines enter the brain and activate the hypothalamic-pituitary-adrenal (HPA) axis; one consequence is that the hypothalamus raises the body's "thermostat." Adjustments then take place throughout the body that elevate its temperature. The most common adjustment is shivering, which generates additional heat by activating many muscles at once. The resulting rise in temperature activates more infection-fighting cells.

As an alternative metaphor to an army, think of the brain and immune and endocrine systems as an orchestra composed of different sections (brass, woodwinds, and percussion). Every orchestra needs a leader, and this role is played in your body by corticotropin-releasing hormone (CRH). CRH neurons in the hypothalamus project to a special docking station of the pituitary called the median eminence. The endings of these neurons stimulate the pituitary to release adrenocorticotropic hormone

One treatment for overstressed minds and bodies, water therapy provides soothing warmth and supportive buoyancy to this Russian war veteran, who suffers from post-traumatic stress.

(ACTH), which sails out into the bloodstream, reaches the adrenal glands, and there stimulates the release of the stress hormone cortisol.

Cortisol not only plays a major role in the stress response but also is a powerful regulator of the inflammatory and immune responses. By inhibiting the production of interleukin-1, cortisol puts a brake on the immune system's tendency to wage all-out war against invaders—a policy that sometimes harms healthy cells and tissues as well as diseased ones. On a second front, CRH also helps modulate the body's inflammatory responses by activating the sympathetic nervous system fibers to influence the thymus, lymph nodes, spleen, and other immune organs; the more activity, the greater the inflammatory response.

The many roles for CRH illustrate a fundamental principle of the brain. Rather than existing in what has been termed "splendid isolation," the brain engages with the immune system in an exquisitely controlled duet. Any disturbance in the HPA axis can result in over- or under-production of cortisol. Too much cortisol shuts off the immune system, raising the risk of infections. Too little pushes the immune system into overdrive and thereby increases the risk for inflammatory diseases like arthritis. Indeed, an increasing number of human diseases are believed to result from impairment of the HPA axis and reduced levels of CRH.

As an example, the puzzling illness known as chronic fatigue syndrome (CFS), and especially its accompanying depression, might be explained on the basis of impaired CRH function. Certainly the symptoms of decreased CRH activity—fatigue,

low energy, vague aches and pains in the muscles and joints—are frequently encountered in CFS with or without depression. And extreme fatigue, sleepiness, and the ingestion of gargantuan amounts of food can mark both illnesses.

CRH may provide a common theme in chronic fatigue syndrome, with too little hormone released in both the arthritic and the depressive components of the illness. Whichever component of CFS is prevalent at a particular time may depend upon whether the cause is physical or psychological. At times, depression may be at the forefront, while the arthritis is quiescent. Later, the arthritis may flare up when the person is feeling generally positive about himself and his situation.

Over the next decade, expect to hear a lot more about treatments for a host of mysterious illnesses like CFS, which are presently misunderstood and difficult to treat. And expect these advances to come from new insights into the intimate connections between the brain and the immune system: the brain-body connection.

In the meantime, much can be gained by learning more about the relationship between stress and mental attitude. While some events are inherently stressful (such as the death of a child), most of the day-to-day things we consider to be stressful are stressful only because of the attitude we take toward them. In short, our stress level is determined by the meaning we give our experiences. For most of us, this may have to be modified in accordance with changing circumstances.

Compare the stress response of our Paleolithic hunter-gatherer ancestors to that of a modern citizen working in an American city. Threats aplenty exist in contemporary society—but they are different from those encountered by our ancestors. Instead of worrying about attacks from wild animals, we fret about attacks from the IRS. Fewer people today worry about contracting deadly infections than they do about how they're going to make their rent or mortgage payments.

———————————

As civilization becomes increasingly complex, changes are required in the balance between the sympathetic and parasympathetic influences. A fight-or-flight strategy becomes less appropriate; today, you almost always have to return and face whatever or whoever it was you were running away from. And, except in self-defense, you get arrested for fighting. Serious health consequences result from these differences in the nature of the contemporary stresses that plague us.

Running from a wild animal mobilizes our body's defenses in an adaptive way. At the sight of the animal, our brain becomes fully alert. Pulse and blood pressure rise. Within seconds, all unessential processes such as digestion are suspended in the interest of mobilizing the fight-or-flight response. Along with physical processes—like pumping more blood to the muscles of the legs and further dilating the respiratory passages—thinking becomes more focused and less susceptible to distraction. The person caught in such a fear-provoking situation will remember the details of the encounter with precision and clarity—an asset in avoiding future fearsome repetitions. Taken together, these steps increase the chances for his survival.

After successfully coping with the threatening animal, the body winds down. Blood pressure normalizes, breathing slows, clarity and alertness

yield to the first stages of the relaxation response. Compare this to the stress responses encountered by many of us today. We often mobilize the same stress response just worrying about the perks that come with our job—or someone else's—or whether our daughter will get into Harvard. Since our brain development allows for what psychologists call increasing cognitive complexity, we tend to mobilize the stress response, according to Stanford University researcher Robert Sapolsky, "for reasons of psychological or social stressors (rather than physical ones). And if you mobilize it for too long, you increase the chances of getting sick. The stress response did not evolve for being chronically activated for reasons of psychological stress."

Of course not everyone has complete control over stressors. Periods of social instability can wreak havoc on mental and physical health. For instance, the collapse of communism preceded increased death and sickness among the citizens of virtually all eastern European countries. Much of the blame probably could be ascribed to higher incidences of heart problems, high blood pressure, suicides, and homicides. In Russia, where political instability was the worst, the life expectancy for men dropped from 64 years to only 59.

As a first step toward eliminating stress in our lives, it's necessary to gauge our own responsiveness to it. Do we go to pieces over things that don't bother other people all that much? A combination of genes and early experiences can affect the brain's wiring, causing some people's "stress thermostat" to be set on high while others remain low. While a total recalibration of that thermostat may not be possible, many things can be done at any age, once we become aware that we are indeed experiencing

The stress response did not evolve for being chronically activated for reasons of psychological stress.

—Robert Sapolsky

stress. Tried and true methods of combatting stress include talking about it to relatives and friends; engaging in some form of physical exercise, such as running or even a long, brisk walk; taking up self-comforting and relaxing activities such as a favorite hobby. Other methods include mental rehearsal and what psychologists refer to as cognitive re-framing.

Suppose that tomorrow is the day of your annual review with the boss. Since things have gone well and your division has increased production, you intend to ask for a raise. The stress of this situation can be significantly decreased by mentally rehearsing positive outcomes of your meeting rather than acting out unproductive mental movies that depict vivid scenes of rejection and embarrassment, following a brusque "No" from the boss. Far better to imagine yourself suggesting the raise and then envisioning your boss nod in agreement.

Such an exercise decreases the stressfulness of your experience and puts you into the relaxed frame of mind that will be required if the boss brings up valid reasons why this might not be the best time for a raise. By thinking about the situation as some-thing you can modify through your input—not as something hopelessly out of your control—you are successfully employing the most helpful stress antidote of all: cognitive re-framing.

If you're sick, replacing stressful foreboding with optimism can save your life. Stress-reduction inter-ventions lower the death rate after heart attacks by 50 percent; stress in hospitalized patients following a heart attack increases the chances of death by three-fold over the next five years and almost doubles the odds of a second heart attack. Clearly,

mental attitude can influence the brain to formulate a healthful or a deadly response.

"Voodoo death"—that is, death from fright—provides another perspective on how attitude can influence the body's responses. Walter Cannon, the physiologist who coined the term homeostasis, was convinced that death could result from a sudden and continuous upsurge in the secretion of epinephrine. He suggested that this hormone flogged the heart into more rapid contractions that eventually led to death. Subsequent tests of this hunch support the view that stress can indeed lead to heart disease. But, at least in test animals, death is even more likely when stress evokes feelings of hopelessness and futility.

In a classic experiment carried out at Johns Hopkins University in the 1950s, rats were put in a water tank from which they had no hope of escape. Some of the animals died within two minutes—while others lasted 40 to 60 hours. Curious as to why rats would show such a range of responses, the experimenters measured the animals' heart rates. Rather than speeding up—as Cannon would have predicted—the heart rates slowed down and finally stopped altogether. In response to an environment filled with hopelessness and futility, it seems, the animals simply gave up.

In the four decades since that experiment, scientists have learned a lot more about the role of hopelessness in illness and disease. Just talking about or describing situations of hopelessness and helplessness can lead to sudden and extreme drops in blood pressure. This can have fatal consequences, either taking the form of sudden heart attacks, or involving the more gradual onset of heart disease. Among men with atherosclerosis—a thickening of blood vessels due to fatty deposits—feelings of hopelessness were associated with an accelerated progression of the disease. Hopelessness also is the best predictor of depression and suicide. If you want a full life, it seems you should—as the song advises—accentuate the positive.

How else might we decrease stress? Perhaps by writing. One group of asthma and rheumatoid arthritis patients spent just 20 minutes a day over three consecutive days, writing about "the most stressful experience" they had ever undergone. Objective tests for disease severity, taken over a four-month period, indicated that nearly half of those in the writing group improved—while only about one out of four of the nonwriters showed improvement. Why should this be?

Writing or talking about a stressful experience objectifies it and lessens the degree of helplessness. If you can't control an illness, you can at least exert some measure of control over your response to it. Writing is merely a more formalized version of something we all do all the time: mental self-talk.

We constantly engage ourselves in internal dialogue, during which we interpret the meaning of whatever is happening to us. It's this *meaning* that we ascribe to the events in our lives that determines our level of stress. Change your attitude now, and the effects will endure into the future.

If you want to cut down on stress, attend first to the things that you're telling yourself. Psychotherapists refer to this form of mental housecleaning as cognitive therapy, and it involves various approaches to cognitive re-framing as discussed above. In a nutshell and minus the mumbo-jumbo, it involves simply eliminating the self-defeating thoughts and behaviors that promote low self-esteem.

Other stress reducers are less cerebral and more physical in emphasis. Fluid and relaxing exercises such as *tai chi* can induce a relaxed frame of mind—I know this from personal experience. But even with exercise, mental attitude is important; an improper approach to exercise actually can worsen rather than reduce stress. Golf and tennis are stress reducers—but only if you can play your best while not worrying about whether you're performing as well as the next guy. And forget altogether about such purported stress reducers as hitting a punching bag while thinking about your boss. Studies have found that such "cathartic" techniques actually *increase* the levels of stress and hostility, since they elicit the same emotions and physiological responses (though to a slightly lesser degree) that would be aroused were you to actually slug your boss.

Remember: Your brain and the rest of your body form a continuum; what you envision in your imagination sets the program for how the rest of your body responds. So think gentle thoughts.

■ FOLLOWING PAGES: Harpists and case workers in Missoula provide palliative care for Mary Bernofsky, a 93-year-old patient suffering from dementia. Nurses report that Mary and others whose illnesses cause them to withdraw from the world can benefit from music and voice, which often provide comfort, if not healing.

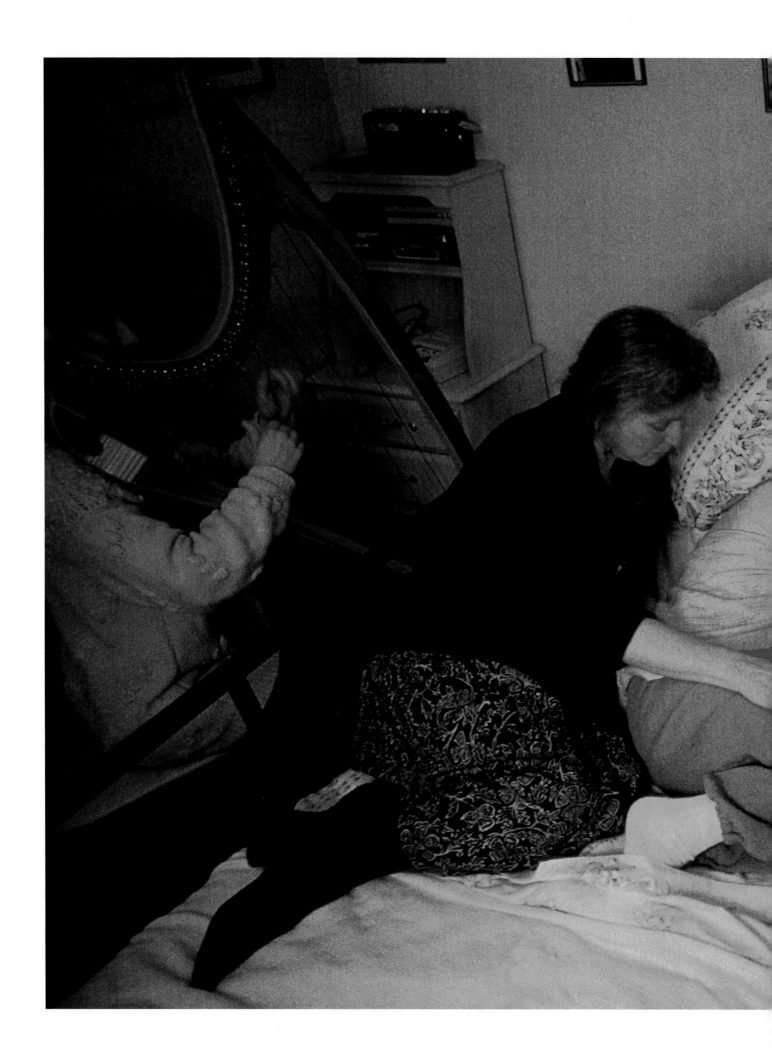

WHEN THINGS GO WRONG

BRAIN DAMAGE

Brain damage is a terrible thing. Not only does it produce crippling physical and mental impairments, but recovery from it is often incomplete or doesn't happen at all.

For instance, until very recently no treatments existed for stroke (a blockage of the brain's blood vessels) other than physical therapy. While the lucky patients achieved some measure of recovery of their paralyzed muscles, the vast majority never returned to the levels of activity they enjoyed before their stroke. Neuroscientists puzzled over this gloomy determinism. Why couldn't the brain—with its vast potential for reorganization among its billions of interconnected cells—recover from a stroke once the blockage was cleared? In order to answer that question, scientists had to delve into the chemical and molecular levels of the brain's organization.

The first clue emerged in 1957, when two researchers accidentally discovered that feeding sodium glutamate to infant mice would destroy neurons in the retinal layer of their eyes. Within a decade, this chance observation was extended by the discovery that glutamate affected neurons not just in the retina but also throughout the brain.

This presented neuroscientists with a major puzzle, because glutamate is the brain's most prevalent neurotransmitter. Normally it is released in precisely controlled quantities that enable one neuron to communicate with another. How could this natural and pervasive brain component also be a destroyer of neurons?

Glutamate, it turns, out, is a bit like the spices in your kitchen: Just a little can work wonders, but too much can spoil everything. When the brain suffers an injury, its injured cells release abnormally

■ Severely brain-damaged, young Mark Finkel responds to a few stimuli, such as playing cards. Although medications can help, many victims of brain injury remain resistant to existing treatments and hope for progress in neural regeneration research.

PREVIOUS PAGES: Long road home: The tremors and unsteady gait associated with Parkinson's disease stem from decreased production of the neurotransmitter dopamine, which results in a general slowing down of the body's motor responses.

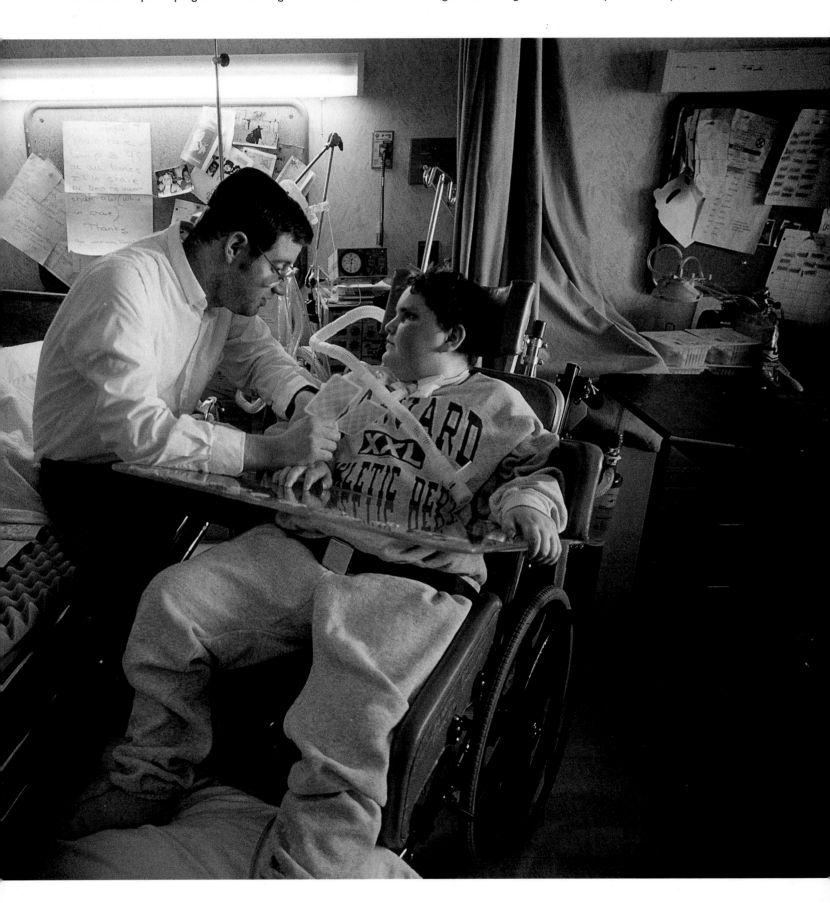

large amounts of glutamate, overstimulating the glutamate receptors on the surface membranes of postsynaptic neurons. This, in turn, causes unusually large amounts of calcium to enter the nerve cells—with disastrous results. Like acid poured onto a computer chip, excess calcium disrupts delicate internal circuits and biochemical processes within the nerve cells. Excess calcium also activates enzymes that produce an especially vicious family of molecules called free radicals.

Free radicals are structurally unbalanced chemicals capable of destroying other compounds that form the internal structure of neurons. As calcium builds up in a nerve cell, it produces free radicals that push the cell's energy-producing structures into meltdown. Eventually the cascade of calcium-induced free radical production leads to the destruction and death of the nerve cell.

Excitotoxicity is the scientific term used to describe the sort of nerve-cell death that results from excessive exposure to glutamate. This process can be triggered by anything that deprives the brain of its normal supply of oxygen. In animal experiments, cutting off the brain's blood supply has been shown to lead to an increased concentration of glutamate in the synapses. This, in turn, causes overstimulation of the receptors on the postsynaptic neuron. Finally, the ensuing flood of calcium into the nerve cell wreaks havoc on the structural integrity of the neuron.

Other experiments have shown that, by administering chemicals that block glutamate's access to its receptors, it's possible to protect susceptible neurons from damage. This technique—called glutamate-receptor blockade—holds promise for improving the outcome for stroke patients. It also has important consequences for treating diseases like hypoglycemia (low blood sugar), traumatic brain injury due to accidents, and *status epilepticus* (a series of back-to-back potentially fatal seizures). Prevention or treatment of excitotoxicity may also prevent or halt the progress of diseases such as Parkinson's, Huntington's, and Alzheimer's.

But glutamate-receptor blockading drugs may need some help, for excitotoxicity is not the only player here. Neuronal death can also result from another mechanism: cell suicide. In response to the damage caused by a lack of oxygen (brought on by stroke or other injury), neurons activate an internal self-destructive chemical program. Treatment involves the timely administration of drugs capable of inhibiting enzymes that promote cell suicide.

So far, treatments involving glutamate-receptor antagonists and anti-suicide agents are limited to animal models of stroke. Will they work as well in humans? "I predict that neuroprotective approaches designed to inhibit excitotoxicity, programmed cell death [cell suicide], or, more likely, both of these mechanisms together will eventually be used in the treatment of stroke patients," according to Dennis Choi, a researcher at Washington University School of Medicine in St. Louis.

Another promising method for brain repair involves stem cells, tiny cells that resemble those of a developing embryo and, like embryonic cells, retain the ability to multiply almost perpetually. Stem cells can give rise to many different types of body cells.

Tissue repair throughout the body depends upon activation and growth of stem cells. Cut your finger, and the healing that occurs is due to the

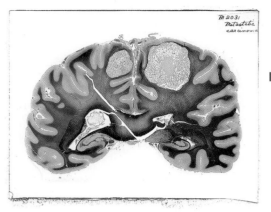

This cross section shows two large metastatic carcinomas—brain tumors— at the top of the cerebrum, caused when cancerous cells elsewhere in the body traveled to the brain and lodged there. Treatment often relies on a combination of surgery, chemotherapy, and radiation.

growth and differentiation of stem cells in the skin. Similarly, the mending of a bone fracture takes place because stem cells transform themselves into the coterie of platelets and red and white blood cells within the healing bone.

For years, most neuroscientists believed that repair within the brain involved mechanisms other than stem cell activation. Primarily, it was thought that surviving neurons responded to brain damage principally by making new connections. But in the 1960s, researchers observed for the first time that the brains of adult rats and other animals could actually generate new neurons. Such neurogenesis occurred in a precise area of the hippocampus called the dentate gyrus.

Over the next two decades, researchers were able to identify new neuronal formation in tree shrews, chickadees, and marmosets. They discovered it also in adult canaries, within those brain areas dedicated to song learning. Learning new songs, in fact, dramatically encouraged canary neurogenesis. Dozens of other experiments showed that, for other species as well, enriched environments and new learning stimulated neurogenesis. If rats were placed in larger cages equipped with the rodent equivalent of toys and exercise machines, they developed bigger and heavier brains. They showed increased cortical thickness in certain brain areas—a marker for additional cells—as well as additional connections between nerve cells. Even more important, adult mice raised under enriched conditions increased the number of new cells in their dentate gyrus by about 60 percent, compared to genetically identical animals raised without perks.

Not only did the animals living in enriched environments grow more cells, they also put those cells to better use. When placed in mazes or in pools of water they quickly extricated themselves. And this enhanced learning curve held true for older mice as well, even though older mice (like people) normally exhibit lower rates of new neuron production. While all of this was very encouraging, neuroscientists wondered whether animal research would ever prove relevant to humans.

In 1998 neuroscientists from Sweden and the Salk Institute for Biological Studies, in La Jolla, announced their discovery that the adult human brain also produces new neurons in at least one site—the dentate gyrus area of the hippocampus.

But just as novelty and environmental enrichment increases neurogenesis, stress reduces it. Remember the effects of stress on the hippocampus? Nerve cells die off and memory erodes. Unfortunately, stem cells also die when stress causes the release of excitatory neurotransmitters in the brain, along with the outpouring of glucocorticoids from the adrenals. Together, these chemicals set off a cascade of destructive reactions within a neuron that culminate in its death.

In essence, such stress-induced damage to adult neurons cancels out any neurogenesis taking place by way of stem cells. So if you want to increase your chances of producing new neurons, keep yourself intellectually challenged and learn efficient ways to manage your stress. Apart from such rather general advice, neuroscientists are uncertain about future applications for stem cell repair in the human brain. Obviously, the creation of new brain cells doesn't usually occur after damage to that organ— that's why recovery often proves so disappointing.

Afflicted with Alzheimer's disease, Art Berliner gets comforted by his wife, Ellen (left). Sometimes, the burden of this illness becomes almost unbearable to Ellen, who cares for him full-time and must seek help for her own emotional pain and physical exhaustion (below).

The nation's most common cause of dementia, Alzheimer's affects as much as 60 percent of all Americans 80 years or older. Although existing medications can ameliorate some of the symptoms and may even slow the progression of the disease, there is no known cure.

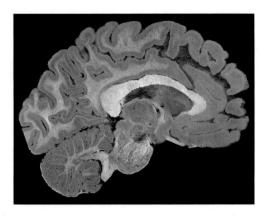

In most instances of injury to the brain, only limited improvement can be expected.

What's needed, according to stem cell researchers Gerd Kempermann and Fred Gage, is the ability "to trace the molecular cascades that lead from a specific stimulus, be it an environmental cue or some internal event, to particular alterations in genetic activity and, in turn, to rises or falls in neurogenesis." Writing in the May 1999 *Scientific American*, they suggest we may soon have much of the information needed "to induce neuronal regeneration at will."

Methods that may spur neurogenesis include the administration of drugs to regulate new cell production, the delivery of gene therapy products to supply needed molecules, transplantation of stem cells, alteration of physical activity, and increasing the challenges to the brain. Researchers admit that only the last two goals are currently achievable.

In the absence of a method for nerve cell regeneration, treatment for brain damage continues to rely on drugs. Researchers hope to find neuro-protective drugs that can halt the progression from localized brain damage to widespread nerve cell death. Creating such agents would have profound implications for the treatment of brain diseases. For example, the blocking of glutamate receptors or the interrupting of cellular suicide could rescue neurons otherwise destined to perish. This would apply both to acute illnesses like stroke, and to decades-long ones, such as Parkinson's disease. The treatments might even exert favorable effects on normal aging of the brain, a process we all must contend with no matter how good our current health.

While aging is not a disease, it is thought to be affected by the same chemical and molecular mechanisms that explain free radical damage. In fact, free radicals head the list of agents presumed to contribute to the aging process. And if aging does result from neurodegeneration caused by an excess of glutamate, then antagonists of glutamate could, at least in theory, halt further degeneration. In order to understand how this might work, let's examine in more detail the glutamate receptor.

Neuropsychiatrist Stephen Stahl, at the University of California at San Diego, considers the excitation caused by glutamate to occur across a broad spectrum. "The spectrum of excitation by glutamate," he believes, "ranges from normal neuro-transmission, to excess neurotransmission causing pathological symptoms such as epilepsy, mania, or panic, to excitotoxicity resulting in minor damage, to slow progressive excitotoxicity resulting in neuronal degeneration such as in Alzheimer's disease, to sudden and catastrophic excitotoxicity causing neurodegeneration as in stroke."

Although not all agree that excitotoxicity plays a major role in panic and mania, the odds are good that the free radicals resulting from the excitotoxic process make major contributions to aging.

Starting in our 20s, our brains begin to shrink. At first the process proceeds imperceptibly, but it gathers momentum with each passing decade. In fact, the process is so obvious that an experienced pathologist performing autopsies doesn't require anything more than naked-eye observations to note brain shrinkage in many elderly people. Traditionally this shrinkage was attributed to widespread nerve cell loss. Observations made decades ago of nerve cell density—the number of neurons in a

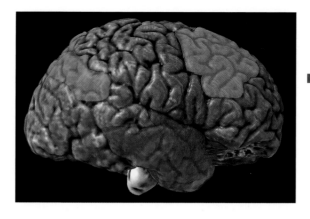

fixed volume of brain tissue—seemed to support this belief. But newer methods for estimating brain cell numbers cast serious doubt on this assumption.

Actual cell counts reveal that the number of cells in the hippocampus doesn't change very much with aging; the same holds true for other cortical areas. Why does the brain shrink? Most neuroscientists believe shrinkage results from changes in the number of glia and other support cells, along with a reduction in the dense tangles of axonal and dendritic branches that connect one neuron to the next.

It's also true that nerve cells beneath the cortex decrease in number, particularly in those centers with ascending projections that fan out to influence the entire brain. Neuroscientist Paul Coleman refers to these subcortical centers as "juice machines," since they distribute the norepinephrine, serotonin, acetylcholine, and dopamine that energize the brain. Due to their widespread distribution, these areas are responsible for such youthful processes as generating mental energy and maintaining alertness. Alterations in these systems can influence youth-associated mental functions like arousal, attention, and motivation.

For instance, an area called the basal forebrain contains acetylcholine-producing neurons that establish contact with many other neurons in the forebrain. Loss of these cells in Alzheimer's disease results in profound memory loss and other cognitive and emotional changes. The same neurons are lost—to a lesser extent—in normal aging. This normal, age-associated loss of neurons also occurs in centers that are responsible for the distribution of norepinephrine, serotonin, and dopamine.

Aging is associated with an increase in MAO-B, one of two forms of the enzyme monoamine oxidase, which breaks down the key neurotransmitters dopamine, serotonin, and norepinephrine. MAO-B generates toxic, brain-damaging free radicals, mainly in the brain stem, limbic system, and frontal cortex. Many of the mental changes associated with aging are associated with changes in these areas.

Current anti-aging efforts focus on reducing the damage done by free radicals. At least one drug for treating Parkinson's disease protects the brain by inhibiting MAO-B and thereby decreasing the production of free radicals. With fewer free radicals, fewer dopamine-producing cells are killed, and the age-associated ravages of Parkinson's are slowed.

Do free radicals exert similar damage in the brains of older people who are free of brain disease? At the moment, the answer seems to be a qualified yes. The current popularity of vitamin E, melatonin, and DHEA as dietary supplements is based on the power of these agents to neutralize the harmful effects of free radicals.

Other compounds may be capable of maintaining and even enhancing brain performance in the older person. In addition to supplements available in

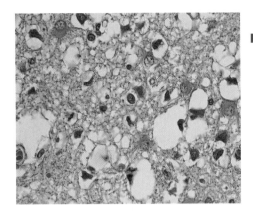

health stores, performance-enhancing drugs known as nootropics (from the Greek for "mind-active") are already facilitating learning and reversing some learning impairments in experimental animals. They are believed to work by altering brain mechanisms considered important in mental performance. For example, piracetam, a derivative of the neurotransmitter GABA, is thought to act as a nootropic by conserving energy reserves and by increasing energy-containing chemicals in the brain. It is one of several so-called "metabolic enhancers" available only outside the United States.

Current FDA regulations restrict pharmaceutical companies to the development and marketing of drugs that treat diseases, rather than enhance normal mental functions. Such regulations hamper further progress in the development of nootropics and, as a result, most drugs with the potential to improve the function of normal brains are being studied as possible treatments for Alzheimer's and other dementias.

For example, a new class of drugs known as ampakines targets one type of receptor for the neurotransmitter glutamate. This receptor is intimately linked with long-term potentiation (LTP), a cellular memory process responsible for changes in the strength of connections between neurons. LTP is believed to underlie long-term memory encoding. The ampakines facilitate LTP and, in both animal and human experiments, enhance memory function.

In one study, 30 elderly volunteers more than doubled the number of nonsense syllables they could memorize after taking an ampakine drug. If these findings are replicated in clinical trials, ampakines may become the first drugs to reverse age-associated memory impairments.

It's important to understand, however, that some mental changes are perfectly normal as we get older. Based on what we know about aging, drugs are likely to remain less important in memory and other cognitive skills than are the enactment of certain lifestyle changes.

As we age, most of us experience increasing difficulty in learning new things—a process that psychologists call fluid intelligence. This lessening of fluid intelligence results from changes in the subcortical areas—the juice machines—that we know contribute significantly to attention and sustained concentration. Grammar, language, and vocabulary don't necessarily decline over time, just our ability to rapidly come up with proper names.

With age we also become more distractable. It takes more effort to keep our thoughts on track and not wander off from our main points. Visual and spatial skills also decrease with aging. Included are some losses in depth perception, spatial localization, and the ability to rapidly identify complex geometric shapes by touch alone. The latter deficit may not make much practical difference in any situation, other than the psychologist's testing laboratory. But loss of depth perception can lead to serious problems in driving a car—one of the reasons some automotive safety experts advise regular retesting of elderly drivers.

Aging isn't all downhill, however. On the positive side, practical experience and problem-solving skills tend to improve as we grow older. Youth, it's been said, is "short on life and short on experience," while maturity is long on both. From that vantage, aging offers benefits that more than make up for the fall-offs in learning new things and rapidly accessing acquired knowledge—especially

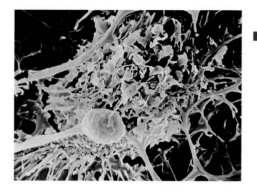

when you consider that many age-associated difficulties can be overcome with practice. The only exception is rapid knowledge retrieval. A younger person will almost always beat an older person when it comes to quickly recalling specific facts or other information (assuming, of course, that each contestant has an equal grasp of the facts). But if you take away the timer, or base the test on world knowledge, older people generally perform at least as well as their younger counterparts.

So instead of dreading the changes associated with aging, we should learn as much as we can about them and look for ways to accommodate ourselves to them. The great pianist Artur Rubenstein provides an inspiring example.

When a long-time admirer commented that the master's skills hadn't changed over more than fifty years of concert performance, Rubenstein paused to correct him. He explained that as he had grown older, he had experienced a slight but noticeable fall-off in his ability to play certain selections to his satisfaction. Rather than becoming discouraged at this turn of events, however, Rubenstein simply altered his repertoire.

If you want to maintain your edge as you age—and improve your chances of living longer—perhaps you should learn to alter your "repertoire" so as to feature your strengths and minimize your weaknesses. Other lifestyle changes likely to prolong longevity include keeping up good physical health by means of regular exercise, eating a sound diet—based on calorie restriction and decreased intake of foods likely to increase free radical reactions, such as polyunsaturated fats—and eating more foods high

in free radical reaction inhibitors, such as fruits and vegetables like cauliflower and carrots.

It's also important to reduce frustrations and problems, by reformulating them into challenges and thereby cutting down on stress-associated brain damage, especially in the area of the hippocampus, where memories are initially encoded.

Finally, and perhaps most importantly, we should keep our brains challenged—by having interesting friends and special interests, by reading new books, and by participating in intellectual challenges such as games, puzzles, and "brain-teasers." The brain is unique among all of the organs in the body. It doesn't wear out, but actually gets better with continued and challenging use. What's more, the brain is dynamic. It's always changing, therefore it doesn't suffer any halfway measures on our part. It either gets sharper and more efficient as a result of our efforts, or it declines in function because we don't ask enough of it.

Thus, in the final analysis, we have no choice but to enhance our brain's performance through activities and interests. When it comes to retaining normal brain function throughout our lives, one inviolate rule reigns supreme: Use it or lose it.

MADNESS, DEPRESSION, AND DISEASE

Imagine a young man who believes Martians are controlling his thoughts by beaming electric currents into his home from outer space. He is anxious and distraught over the feelings aroused by this bizarre conviction. A few decades ago such a person would have been confined to a mental hospital. No treatment was available and improvement was a hit-or-miss proposition. But in the mid-1950s antipsychotic drugs became available, eliminating the need for hospitalization and, in some instances, clearing away delusions.

At that time—after more than 30 years of research—neuroscientists believed dopamine was the main culprit in psychotic reactions such as schizophrenia. This illness is marked by delusions (the Martians), hallucinations, disorganized speech and behavior, blunted effect (emotions), social withdrawal, and other indications of disturbed thinking and feeling. Although a biological explanation for schizophrenia remains incomplete, drugs that increase dopamine in key synapses in the brain are known to worsen the illness, while drugs that decrease dopamine within those synapses diminish or halt many symptoms of the disease.

For instance, cocaine and amphetamine both worsen schizophrenia. This is because both of these powerful mind-altering substances increase the

■ When no one cares: Abandoned to the state, a boy at the Institution for Social Care of Children, in the former Czechoslovakia, screams alone. He is one of many children from a heavily polluted area who were born with mental retardation, paralysis, or other diseases of the brain.

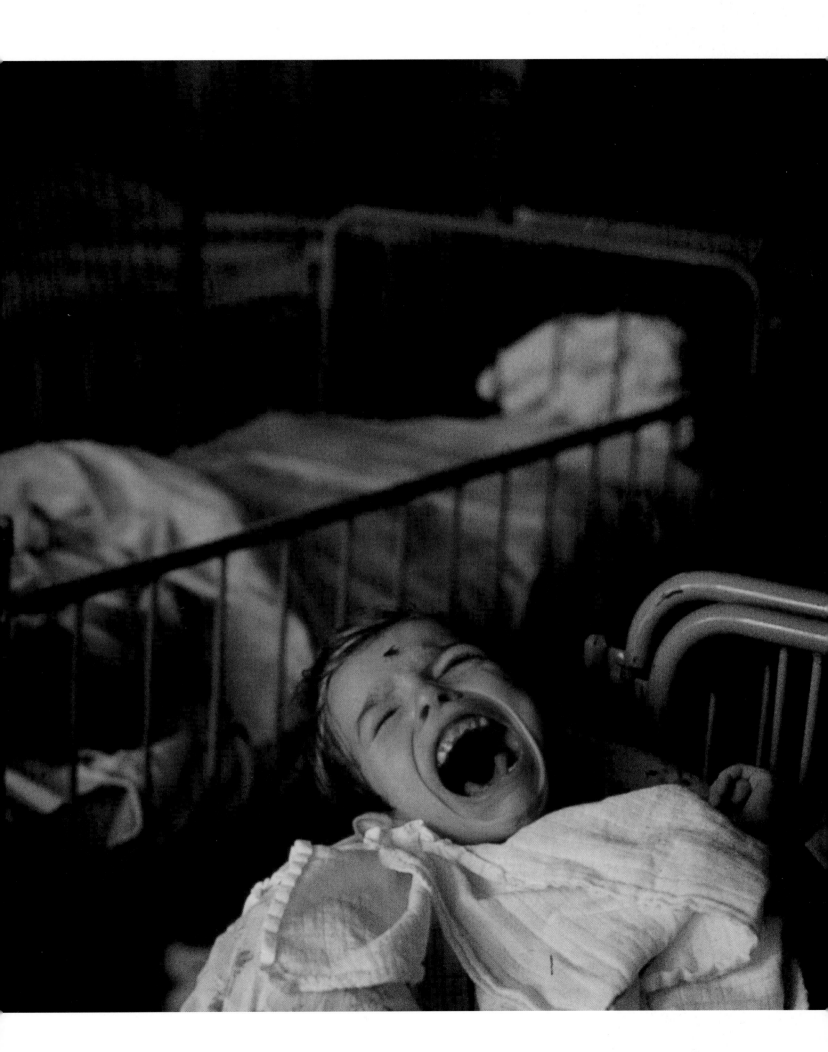

release of dopamine. As the level of dopamine rises in the synapses within the limbic system and elsewhere, the symptoms of schizophrenia worsen. Even in people with no previous signs of mental illness, amphetamine and sometimes cocaine can induce a paranoid psychosis that experienced psychiatrists can't always distinguish from the initial onset of a classic case of schizophrenia.

Antipsychotic drugs act by blocking dopamine receptors, especially a subclass known as the D2 receptors. There are four dopamine pathways fanning upward from the midbrain. One of them, the mesolimbic, extends into the limbic system, which is involved in the production of delusions and hallucinations. When this pathway is overactive, more dopamine is released at the synapse, bringing on the symptoms of psychoses such as the Martian delusion. Just the opposite happens when an antipsychotic drug occupies the dopamine receptors, blocking the neurotransmitter from bonding: The bizarre conviction about Martians attenuates or may even completely disappear.

But the drug may also affect the other three dopamine pathways, causing various side effects. Occupation of the nigrostriatal pathway, for example, can lead to Parkinson's-like symptoms, because Parkinson's disease stems from a loss of dopamine-producing cells within this pathway. That's why a schizophrenic on dopamine-blocking drugs may move slowly or become tremulous.

Worse yet is tardive dyskinesia, also related to the nigrostriatal pathway. This disorder is marked by rapid and irregular movements of the face, tongue, neck, arms, and legs. Symptoms can include lip-smacking, chewing, tongue protrusions, facial grimacing, and sudden and irregular movement of the limbs. These disturbing and sometimes irreversible actions are side effects, caused by using an antipsychotic drug that unfortunately alters the dopamine receptor. Such examples remind us that chemical manipulation of the brain—even with the best intentions—can sometimes result in unexpected and even disastrous consequences.

One way to avoid the extremes of too much or too little dopamine is to hit only selected targets in the dopamine circuit. Some of the more recent antipsychotic drugs do this, principally influencing only those dopamine pathways and receptors that are involved in the production of a particular psychosis. Thus pathway two (the mesolimbic) and, to a lesser extent, pathway three (the mesocortical) are often targeted. But such efforts at biochemical precision are complicated by the fact that neuroscientists don't completely understand the biological basis of the illness.

For patients with schizophrenic psychosis, the current goal is to control and reduce their symptoms, not cure them. While this may seem overly modest, consider diabetes and hypertension: They aren't cured by medications, either; they're controlled. In much of medical science, control rather than cure actually is the rule. That's certainly the case with many neuropsychiatric illnesses. Whatever treatment a patient undergoes, no one can ever guarantee that the panic attacks, depression, or psychosis will never return. This is especially important to keep in mind when considering some of the more common emotional disturbances, such as anxiety.

Today, anxiety has become the most frequently encountered mental affliction in the world. We are fortunate in that neuroscientists already know a lot about the neurochemical basis of anxiety and the chemical antidotes that can relieve it. The first breakthrough came about in the 1950s, when researchers reported that the drug Librium exerted a "taming" effect on agitated monkeys. Could this effect be extended to humans and thereby calm anxious patients? That question spurred a whole generation of psychopharmacologists to come up with drugs known as the benzodiazepines, the first class of tranquilizers with widespread application in the treatment of anxiety.

Since then, psychopharmacologists have developed numerous benzodiazepine tranquilizers, including Valium (the "Mother's Little Helper" referred to by the Rolling Stones in a song by that name), also Ativan, Xanax, and Klonopin. In fact, the benzodiazepines have become the main treatment for anxiety disorders, particularly the common one known simply as generalized anxiety disorder. They work by binding to specialized receptors within the limbic structures—especially those in the amygdala and hippocampus—in addition to other sites in the brain stem and cerebral cortex. After treatment with these drugs, 65 to 70 percent of patients experience significant improvement.

The discovery of such sites raises an intriguing question: Why would nature endow the brain with selective sites for binding to a chemical that was invented only in the 1950s? The most likely explanation, of course, is that the brain contains *natural* benzodiazepine-like substances, which serve as nature's "little helpers" by docking at those specific sites. Presumably, that's exactly what happens when we calm ourselves without the help of medicines. Because man-made benzodiazepines can activate these same receptors, they effectively mimic that natural, self-generated feeling of tranquility.

We now know that such tranquilizers work essentially by acting on receptors that strengthen the effects of the inhibitory neurotransmitter GABA (gamma aminobutyric acid). This enhances GABA's inhibitory action and leads to a decrease in neuronal firing.

On the molecular level, what we call "anxiety" is basically a diffuse overactivity of certain receptor neurons. Tranquilizers oppose this overactivity by enhancing the inhibition of those neurons and stabilizing their membranes. As neuronal activity is reduced, the person's corresponding anxiety gives way to a more tranquil mood.

The amygdala, the limbic structure associated with the emotional aspects of memory, is a prime target for anxiety reduction, in part because it is especially rich in GABA receptors. And, since serotonin-enhancing drugs also enhance GABA, they increase neuronal inhibition. As a result, the amygdala becomes less active; conditioned fear becomes harder to establish. An amygdala-inhibited rat will no longer react to sounds or odors, only to stronger stimuli such as an electric shock. In humans, similar amygdalar inhibition leads to a lessening of the symptoms of anxiety.

Another strategy for pacifying the amygdala involves the administration of calcium-channel blockers. As described earlier, the establishment of memory is dependent upon the entrance of calcium into neurons. Long-term potentiation, a process that relates to memory at the cellular level, can be blocked by chelators—drugs that bind to calcium and thereby prevent its entrance into the neuron.

The amygdala can also be tamed by drugs known as GABA agonists, which exert the same effect as the natural neurotransmitter. By reducing amygdala activity, they also reduce fear-conditioned learning. The process is similar to one treatment for Parkinson's disease, which involves the administration of agonist drugs that occupy and stimulate the same receptors as natural dopamine. This results in improvement in the symptoms of the disease.

Other drugs that work on the benzodiazepine receptor can actually worsen anxiety, however, for they block the tranquilizing effects of benzodiazepine. The result: increased agitation and signs of overstimulation of the autonomic nervous system, such as increased heart rate, blood pressure, and breathing. To some neuroscientists, such anxiety-inducing effects of certain drugs suggest the existence of naturally occurring compounds in the brain that might be responsible for the generation of anxiety. Such compounds—if they do exist—probably differ in amount from person to person and may even determine each person's anxiety "set-point."

Anxiety-inducing substances exert just the opposite effect that GABA does. Alcohol, for instance, interferes with GABA's inhibitory actions. This explains the behavioral disinhibition produced by alcohol, such as the "bad drunk" syndrome, which includes loudness and lack of self-control. Later, as the effect of alcohol wears off, the GABA receptor becomes unstable, sometimes escalating to the extremely anxious responses that can accompany alcohol withdrawal. Since benzodiazepines and alcohol affect the same GABA receptor, they exhibit

what's called cross-tolerance. Tolerance means that, over time, the same amount of drug produces less effect; cross-tolerance means that exposure to one drug leads to increased tolerance to a second drug. For this reason, an alcoholic may remain awake after consuming amounts of benzodiazepine drugs that would put an ordinary person into a coma. One advantage of cross-tolerance, however, is that it can form the basis for detoxification treatments for alcoholism. The process involves substituting a benzodiazepine for alcohol and then slowly tapering the drug to minimize withdrawal effects.

But in instances of benzodiazepine overdose—which are rare—treatment must proceed at a quicker pace, because high amounts of these drugs interfere with breathing and normal heart rate. Such effects can be reversed if the emergency room doctor acts quickly enough and administers a potent antagonist to block the actions of the drug at its receptor, which causes the patient to suddenly awaken from his benzodiazepine-induced coma.

It's important not to ignore anxiety because, if left untreated, it has the potential to alter the brain irreversibly. James G. Barbee, director of Louisiana State University's Anxiety and Mood Disorder Clinic, warns that anxiety disorders "may be due to initial changes in neurotransmitter activity, which ultimately may lead to changes in gene expression and the structure and function of cells in the brain." Fortunately, we can treat various types of anxiety.

One form of anxiety disorder that is treatable is known as obsessive-compulsive disorder, or OCD. An obsession is a persistent, intrusive impulse or idea that produces anxiety and distress. For instance,

Photographed in 1886, this inmate of the Imbecile Asylum in Burlington, New Jersey, wears an expression that might sum up the 19th-century "solution" for the mentally ill: Institutionalize them in human warehouses and offer little in the way of real treatment for their diseases.

a religious person may be tortured by a persistent idea that he may set fire to his church. A compulsion is an attempt to get rid of the obsession-generated anxiety by some form of repetitive behavior. In this instance, the person with the obsession about starting a fire may obtain momentary relief from his recurring thought by compulsively eliminating all the matches or other flammable items from his home. Compulsions are often very elaborate, time-consuming, and psychologically costly mechanisms for temporarily reducing the anxiety that can be generated by an obsession.

For years, psychiatrists prescribed psychological treatments for OCD, often relying on psychoanalysis or other "talking therapies." But biologically oriented doctors remained confident that the illness resulted from physical rather than psychological causes. Neurologists provided some proof for this in the 1920s, when they observed that obsessions and compulsions, along with the accompanying anxiety, can result from brain injury.

Today, thanks to PET scans, the anatomy of obsessive-compulsive disease is well understood. Hyperactivity in part of the frontal lobe, the anterior cingulate, and part of the basal ganglia (the caudate nucleus) accompany obsessive-compulsive episodes. Activity in these areas decreases between episodes after successful treatment.

In one famous experiment, a man obsessed with cleanliness forced himself while in the PET scanner to touch "soiled" and "clean" towels (actually both towels were clean, but he could not see them). When handling the "soiled" towel, PET scan activity increased in his orbitofrontal-anterior cingulate-caudate circuit. When handling the "clean" towel, activity in these areas plummeted. This dramatic

correlation suggests that hyperactivity within these brain circuits may serve as a marker for obsessive-compulsive illness. In the not-too-distant future, PET scans may be used to confirm OCD diagnosis and to monitor the success of the treatment.

Treatment of OCD usually involves drugs that increase the concentration of serotonin within key synapses of the brain. One class of inhibitors increases synaptic serotonin by blocking that neurotransmitter's normal movement out of the synapse and back into the presynaptic neuron.

Drug treatment for OCD works best when combined with psychotherapy that concentrates on correcting distortions and inappropriate assumptions in the patient's thinking. In the case of the fire-obsessed OCD sufferer, a therapist might begin by talking about all of the positive effects of fire and matches. In time, the patient will be encouraged to handle matches and even light the logs in the fireplace, thus reassuring him that such activities do not necessarily lead to arson.

The key to treatment, according to Gail Steketee, a professor at Boston University's School of Social Work, is to break the link between obsessions that increase anxiety and compulsions that reduce it. "The goal of treatment," she maintains, "is to stop compulsions so that the patient can become used to intrusive obsessive thoughts."

While either behavioral psychotherapy or serotonin-like agents alone can successfully bring about patient improvement in OCD and other forms of anxiety, many clinicians agree that a combination of the two methods offers the best chance for a permanent cure.

■ Time passes slowly for this schizophrenic prisoner in a Nashville jail.
Distinguishing criminality from mental
illness is often difficult, especially
when the offense involves explosive
outbursts of violence or a hostile
response. Just such an act can be
the first indication of brain disease.

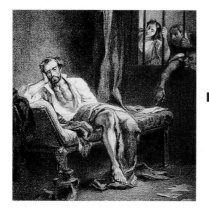

One of the most common mental diseases of humankind, depression probably afflicts this 16th-century inmate of the Asylum of St. Anna, in Ferrara, Italy—while visitors bestow one of history's all-too-common responses to the deranged: mockery.

Second only to anxiety, depression ranks as one of the most common mental disorders afflicting our species. Yet while everyone has been anxious at some or another, not everyone has experienced true depression. Garden-variety low moods don't qualify; it's perfectly normal to feel a bit down, now and again. Nor is normal bereavement at the death of a loved one the same as depression. Bereavement and loss are painful but normal processes we all must undergo on some occasions during our lives. Yet I have encountered in my professional practice many mourners asking for antidepressants to help them get through a funeral or memorial. In almost all instances I gently refuse such requests, telling them that to reach for an antidepressant at such a time is to misuse the drug.

The distinction between true depression and normal day-to-day setbacks in mood is important, because recurrent depressive episodes are treatable and even preventable, thanks to currently available medications. According to the Diagnostic and Statistical Manual of Mental Disorders, a major depressive episode includes several of the following:

• Depressed mood most of the day, nearly every day, as indicated either by the patient or through observation by others.
• Markedly diminished interest or pleasure in all or almost all activities most of the day, nearly every day.
• Significant weight loss or weight gain, amounting to a change of more than 5 percent of body weight in a month. Significant decrease or increase in appetite nearly every day.
• Insomnia or hypersomnia (drowsiness) nearly every day.
• Psychomotor agitation (agitated thoughts or restless pacing) or retardation (slowed thoughts and movements) nearly every day, observable by others—not merely one's subjective feelings of restlessness or of slowing down.
• Fatigue or loss of energy nearly every day.
• Feelings of worthlessness or of excessive or inappropriate guilt (which may be delusional) nearly every day.
• Diminished ability to think or concentrate, or heightened indecisiveness, nearly every day.
• Recurrent thoughts of death or recurrent suicidal thoughts, without a specific plan for suicide or an attempt at carrying it out.

In the early years of psychopharmacology—the treatment of emotional illness with drugs rather than psychotherapy—researchers thought that depression resulted from a disturbance in a single receptor or class of receptors. An early, oversimplified theory known as the monoamine hypothesis held that depression generally resulted from a lack of norepinephrine, dopamine, or serotonin.

According to this theory, drugs that increased the synaptic levels of these monoamine neurotransmitters exerted an antidepressant effect. Just the opposite occurred with drugs that depleted the synapses of these neurotransmitters.

Today, neuroscientists realize that depression is far more complicated. They no longer believe that it results simply from disturbances in a single neurotransmitter, or even from imbalances in a family of chemically similar neurotransmitters. When it comes to depression, it's best to think of all neurotransmitters as separate instruments in a symphony. Just as a symphonic performance can be

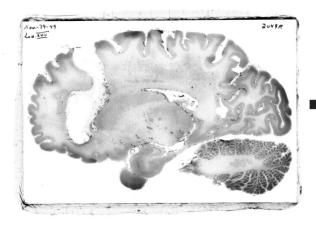

Once considered a reasonable treatment for a variety of mental ills, frontal lobotomy is an extreme surgical procedure that destroys specific parts of the brain, as seen here. Modern alternatives include drugs that modify thoughts, moods, and behavior.

spoiled by failures on the part of any one instrument, so, too, depression can occur as a result of disturbances of any one of many neurotransmitters.

The earliest pharmacological antidepressants, introduced in the mid-1950s, consisted of two classes of agents. The first, known as the tricyclics, was so named because drugs in this class contained a chemical structure comprised of three rings.

Normally, after norepinephrine, serotonin, or dopamine bind to their respective receptors, they are released and transported back into the presynaptic neuron. But this process can be blocked if an antidepressant drug binds to the neurotransmitter's transporter, preventing it from linking to the neurotransmitter and carrying it back to the presynaptic neuron. As a consequence, the neurotransmitter remains within the synapse. It's this synaptic buildup of the neurotransmitter that exerts the antidepressant effect.

In essence, that's how the tricyclic drugs work; they block the transporter molecules or "pumps" that ordinarily sweep dopamine, norepinephrine, or serotonin out of the synapse and back to the presynaptic neuron. The neurotransmitter accumulates in the synapse, and, for reasons that no one is entirely certain about, this buildup brings about an antidepressant effect.

The patient's depression doesn't respond at once, however. The antidepressant drug usually induces chemical alterations in the brain several weeks before any improvement in the patient's depression. Thus, while the particular neurotransmitter imbalance may be corrected within days, the patient's recovery from depression usually takes

weeks or even months. Why the delay? And what happens during the interval between administration of the drug and improvement of the patient?

For one thing, antidepressants, whatever their initial effect on one or more neurotransmitters, relieve depression by altering the *balance* of neurotransmitters within the synapse. This takes time. Returning to the symphony analogy, each musical composition dictates a rebalancing of all the instruments; a featured soloist in one composition often provides only background accompaniment in another. When the depressed person responds to an antidepressant drug, it's as if the neurotransmitters—the mind's musical instruments—are brought into a new balance with one another. Some that previously played leading roles now are relegated to supporting roles, and vice versa.

Tricyclic antidepressants work by altering the balance of serotonin and norepinephrine, along with exerting some weaker effects on dopamine. In fact, a typical tricyclic is actually five drugs in one. Its norepinephrine and serotonin portions exert a positive antidepressant effect, while the other three portions blockade receptors for acetylcholine, histamine, and a third compound. This blockading action is far less desirable, for it causes annoying and sometimes troublesome side effects. Activation of the histamine portion often produces weight gain and drowsiness. Activation of the cholinergic portion leads to dry mouth, constipation, and blurred vision. Activation of the third antagonist results in dizziness and low blood pressure.

Unfortunately these side effects can occur with the drug's first dose—while the antidepressant effects take several weeks to "kick in." Some patients are so put off by the prospect of going about their day drowsy, constipated, and cotton-mouthed that they stop taking tricyclics before their depression has completely responded.

A second class of antidepressants, known as monoamine oxidase inhibitors or MAOIs, employs a different mechanism to increase the amount of neurotransmitter in the synapse. They inhibit the action of an enzyme (monoamine oxidase) that breaks down norepinephrine, dopamine, and serotonin into their chemical components. By inhibiting this enzyme, the effect of the neurotransmitters is prolonged, thus relieving the person's depression.

Although tricyclics and MAOIs work by different methods, early psychopharmacologists sometimes prescribed both for the same patient, in an attempt to double the chances of success. But doing so is dangerous. The combined effect of these two different classes of drug can result in too much serotonin in the synapse, which can lead to a toxic reaction and even death.

Today, selective serotonin reuptake inhibitors, or SSRIs, have largely displaced both the tricyclic and the MAOIs as the treatment of choice for depression. They have fewer side effects and can be taken in large amounts without fear of a lethal outcome. As their name implies, they act principally on serotonin, inhibiting the reuptake of that neuro-transmitter from the synapse, thus leading to a higher concentrations of it. By and large, they do not affect the transport mechanisms for dopamine or norepinephrine. Prozac, at one time the most heavily prescribed drug in this class, now shares the stage with other SSRIs such as Paxil and Zoloft, which also target serotonin.

Other recently introduced antidepressants such as Wellbutron, or bupropion, also enhance serotonin levels—and take aim at norepinephrine and dopamine as well, by blocking their transporters. In addition, some new antidepressants such as Effexor, or venlafaxine, block the reuptake of all three of these neurotransmitters.

———————

With so many drugs and so many mechanisms of action available, a question arises: Why not just one drug for all? For one thing, neuropsychiatrists are convinced that the underlying chemistry of depression isn't the same for everybody. Certainly the symptoms vary from person to person. Some depressed people cry, wring their hands, and pace about in a state of internal misery. Others lose interest in events around them, become reclusive, and may even deny feeling particularly "low."

Chemical variations seem likely within the brains of people experiencing different varieties of depression. Norepinephrine may be the chief culprit in one form of depression, while serotonin may play the major role in another. Because of this chemical variability from person to person, drugs with differing modes of action make sense. Whatever the chemical changes associated with a particular individual's depression, some alteration at the synapse will be required in order to treat it.

The existence of so many antidepressants with so many different mechanisms of action also serves as a reminder that the brain mechanisms underlying depression aren't fully understood.

"In fairness, it must be stated that we really do not have a complete and adequate understanding of how antidepressant drugs work," writes Stephen M. Stahl, a neuropsychiatrist and psychopharmacologist. For one thing, as mentioned, antidepressant drugs exert immediate and measurable effects within the body—but several weeks may pass before they exert an antidepressant effect. The solution to this puzzle can be found in the chemical structure of the particular antidepressant.

While antidepressants don't cure depression, they can control it so well that longevity and general health need not be affected. Untreated depressives have a high suicide rate. Will future antidepressants eliminate depression as a major cause of death and disability? Will medications ever provide a cure for depression? Any answer must take into account the complexity and deeply mysterious nature of this illness. Many people undergo experiences that might be expected to result in depression, yet remain resilient and reasonably upbeat. Others become depressed for no apparent reason.

In the book *Darkness Visible, A Memoir on Madness*, author William Styron details his own depressive illness. He calls depression "A true wimp of a word for such a major illness." It is, he points out, "a noun with a bland tonality and lacking any magisterial presence, used indifferently to describe an economic decline or a rut in the ground."

At the moment, the many interactions that occur between neurotransmitters and their receptors offer our best hope of finding satisfactory treatment for depression. Additional insights into the chemistry of depression will come, and they should prove particularly helpful in furthering our understanding, control, and treatment of stress.

Although stress is an unavoidable aspect of modern life, we can in most instances modify and control our responses by transforming stresses into challenges. But not always. The latest neurobiological findings indicate that some stresses, especially if undergone early in life, can lead to a lifetime of physical and psychological impairment.

Freud anticipated such findings, emphasizing that early infancy and childhood experiences could shift the emotional balance towards serious psychiatric illnesses. For instance, early child abuse or neglect increases the likelihood of depression in adulthood. An experiment carried out on rats provided clues to the mechanism involved.

Shortly after birth, one rat was separated from its mother for a three-hour period every day for two weeks. As a result of this enforced separation, the mother-pup relationship became seriously and permanently disturbed. The mother treated her pup with indifference, even neglect. Tests of blood and spinal fluid from the pup suggested a probable mechanism: The release of stress hormones had put the neglected young in a "survival mode," in which it had conserved food and energy at the expense of its normal development and growth.

Stress, of course, causes the hypothalamus to instruct the pituitary to put out more ACTH, or adrenocorticotropin hormone. It accomplishes this by secreting another hormone, corticotropin releasing factor (CRF), which travels via a special arterial pathway to the anterior pituitary. Once released into general circulation, ACTH wends its way to the adrenals, the endocrine glands that sit atop the kidneys. The adrenals—the foot soldiers

■ FOLLOWING PAGES: One form of therapy: Mentally disturbed adults find a productive outlet in a Pakistani carpet-weaving factory.

in this operation—respond by producing cortisol and other stress hormones.

Normally, a rising level of cortisol in the blood acts as a feedback mechanism to reduce production of ACTH by the pituitary. But under conditions of uncontrolled stress—such as the separation of rat pups from their mothers, or severe depression in a human—this feedback loop fails. A vicious cycle ensues: Excess CRF is secreted, driving the pituitary to release more ACTH, which in turn drives the adrenals to put out more cortisol. High levels of ACTH and cortisol are the ingredients for a full-blown and sustained stress response.

Following the isolation and genetic sequencing of CRF in the early 1980s, neuroscientists were able for the first time to map its centers of activity. According to Charles B. Nemeroff, a research psychiatrist at Emory University, in Atlanta, CRF "in addition to its influence on the hypothalamic-pituitary-adrenal (HPA) axis projects to two centers important in mood and anxiety: the amygdala, and the locus ceruleus, the area of origination of 70 percent of the brain's norepinephrine." Such projections strongly indicated an important role for CRF in depression and anxiety disorders.

When injected into the brains of laboratory animals, CRF induces responses similar to human depression: The animals eat less, their sleep is disrupted, they lose interest in sex, and they react with fear when faced with new situations, such as a change in living conditions. Rhesus monkeys react to CRF injections with what primate research pioneer Harry Harlow termed "behavioral despair syndrome"—they crouch in the corner of their

cages, rock back and forth, and avoid interaction with other monkeys.

Based on such animal findings, Nemeroff and others began work on a CRF theory of depression. They examined samples of spinal fluid from untreated depressed patients, and found they contained twice the normal concentration of CRF. They also found increased CRF production in the hypothalamus and other brain areas that don't normally express CRF. Most intriguing of all, depressed women with a history of sexual abuse in childhood or adolescence showed an increased response to stress along the HPA axis.

"These results suggest that alterations in CRF neurons in the hypothalamus mediate the effects of early trauma and increase an individual's vulnerability to depression," observed Nemeroff.

Since CRF plays such a prominent role in stress and the depression that frequently accompanies it, new antidepressants are in production that will block CRF receptors in the brain. If successful, these drugs will alter abnormal responses to stress and thus should help increase individual tolerance to stress among people who have experienced abuse or neglect early in life. At least that is the hope of the researchers.

Another severe form of stress intolerance, post-traumatic stress disorder (PTSD), develops later in life but with equally disabling consequences. One PTSD authority sums up the symptoms: "Traumatic experiences scar the traumatized individuals, weakening their resilience to future stress.... It appears that even when...post-traumatic stress disorder remits, or evolves into a more stable form, the afflicted person may become highly sensitized to stress in general. He is permanently altered, harboring the potential for future response in re-exposure to threatening stimuli." Although this description captures the essence of PTSD, it doesn't go far enough in defining the "threatening stimuli."

Most of us don't freak out when somebody drops a dish, or when the growl of thunder interrupts a pleasant summer afternoon with the promise of an impending storm. But a person suffering from PTSD experiences those events differently. If he's a military veteran, those sounds may evoke the vivid sensations associated with combat; he might experience panic or explode in a rage. That's because either the dish or the thunder incident activates a hair-trigger startle response in him.

───────────

In the laboratory, panic attacks can be triggered in persons afflicted with PTSD. To do this it's only necessary to administer the drug yohimbine, an alkaloid derived from a tropical African tree. Yohimbine enhances the activity of norepinephrine— a key component of the fight-or-flight response— by blocking its receptor. Neurons that manufacture norepinephrine attempt to overcome the receptor blockade by producing more and more of the neurotransmitter. This translates into an increased firing rate of these neurons, and more norepinephrine being released per impulse. This, in turn, causes many of the symptoms of PTSD: Racing heart, rapid breathing, sweating, and a sense of impending dissolution and terror.

Interestingly, the administration of a drug that interacts with serotonin receptors can also set off PTSD-associated panics by altering the balance of serotonin to other neurotransmitters, especially epinephrine. Moreover, people who experience

attacks in this situation fail to develop panic when given yohimbine. These findings suggest that PTSD can result from disturbances in more than one neurotransmitter.

Dennis Charney, a Yale University psychiatrist who also serves as director of the National Center for Post-Traumatic Stress Disorder, states, "The neurobiological consequences of exposure to severe psychological stress involve alterations in a number of neurotransmitters, neuropeptides and amino acid transmitters. These changes consequently produce dysfunction in neural circuits that mediate the complex symptomatology associated with post-traumatic stress disorder."

If more than one neurotransmitter is involved in PTSD, it makes sense to assume that drugs that influence those neurotransmitters may provide effective treatments for PTSD. For instance, if norepinephrine encodes a memory of a traumatic event within the brain, then a blocker of the norepinephrine system may prevent encoding of that traumatic memory. (As mentioned in the chapter on memory, if encoding doesn't occur, subsequent recall can't take place.) Under certain circumstances, this medication-induced "amnesia" could prove beneficial. Consider, for example, how helpful such a drug might have been, had it been available for those horrified students rushing out of Columbine High School in April 1999, during the worst school-associated mass murder in U.S. history.

For now, however, the most promising approach to PTSD involves treating victims with antidepressants that affect either norepinephrine or serotonin receptors, or both. One drug still awaiting approval, reboxetine, works by targeting the norepinephrine system. It may prove useful not only after a traumatic experience like Columbine, but also as a prophylactic agent. Police and rescue workers might take it in advance and thus be better prepared to deal with particularly horrific crime scenes. The rationale for such a chemical preemptive strike is well established. Stress researchers have learned from bloody tragedies like the 1995 bomb explosion in Oklahoma City that rescue workers are at very high risk of developing PTSD.

Drugs that block the reuptake of serotonin from the synapse (that is, SSRIs) have already been shown to provide a reduction in PTSD symptoms although not complete remission. But when dealing with such a disabling and psychically anguishing illness as this, just lessening the symptoms can mean the difference between a functioning, reasonably happy individual and a psychiatric casualty.

Yet another approach to PTSD involves the use of CRF-receptor antagonists, which may work as prophylactic agents in the prevention of depression and anxiety. Such compounds, researcher Charles Nemeroff believes, have the potential to "prevent the development of PTSD in rape victims."

There is an additional and vitally important reason why antidepressants should be employed: PTSD causes brain damage. Specifically, the total number of neurons is reduced in the hippocampi of people afflicted with PTSD. This loss of brain cells results in a reduced hippocampal volume (revealed by MRI scans). In soldiers, these changes vary directly with the amount of combat exposure: The greater the exposure to fighting, the greater the loss of hippocampal cells. I find it extremely interesting that a decrease in hippocampal volume

is often observed among people who have been subjected to abuse in early childhood.

If continued long enough, treatment with antidepressant drugs results in the production of new hippocampal cells, at least in laboratory rats. Does the same hold true for humans who have been exposed to the extreme stresses that lead to PTSD? Nobody knows for sure yet, but many specialists believe that the chances are good that antidepressants have some beneficial effect on the size as well as the function of the hippocampus. This could either take the form of more cells or more efficient communication among existing cells. In either case, memory should improve, along with many of the symptoms that render PTSD so disabling.

Fortunately, most of us are spared the kinds of horrific experiences that lead to the development of PTSD. Even so, the findings on the neurobiology of stress, anxiety, and depression can pay off in terms of the understanding and treatment of other and more common neuropsychiatric illnesses.

For instance, migraine headache, a disabling condition that afflicts between 16 and 18 million Americans, provides valuable insights into the ways a brain illness can be understood and conquered through improved understanding of the interaction of neurotransmitters and their receptors.

While headaches are their most common symptoms, migraine sufferers also experience mood swings, dizziness, intolerance to bright lights or loud sounds, disturbed thinking, and impaired concentration. Many experience depression as well—and many depressed people suffer migraines. This overlap in symptoms suggests that similar neurotransmitters may be involved in the two

> The neurobiological consequences of exposure
> to severe psychological stress involve alterations
> in a number of neurotransmitters, neuropeptides
> and amino acid transmitters.
>
> —Dennis Charney

conditions and that findings about one may provide valuable insights into the other.

Treatments for headaches have varied widely through history. The ancient Egyptians strapped a clay crocodile containing sacred grain to the head of the sufferer with a strip of fine linen engraved with the name of a god. Another treatment, trephination, was more direct and far more dangerous: It consisted of boring holes into particular areas of the skull, in hopes of releasing malevolent spirits or pressures believed to cause the ache. Practiced to varying degrees by many civilizations throughout the world, trephination persisted in Europe as late as the mid-17th century—and in fact still has some adherents.

———————

Yet other treatments for migraine have involved opium, hemlock, sneezing powders, even the removal of a pint or two of blood. They have varied so widely and so bizarrely simply because no one really knew the cause of the affliction. Followers of Hippocrates attributed migraine to an imbalance of the four elements—phlegm, blood, black bile, and yellow bile. In the Middle Ages, some blamed migraine attacks on demonic influences. The 17th-century English anatomist and physician Thomas Willis ascribed headache pains to the products of widely dilated blood vessels in the head.

In the early 20th century, medical doctors correctly reasoned that migraine was somehow related to disturbances in the arteries feeding the scalp and brain. They weren't sure exactly how. Could anything be done to stop the process and abort the dreaded migraine attacks?

By the 1960s, doctors noted that boosting the level of serotonin in the bloodstream decreased the symptoms of migraine. This observation was followed years later by the discovery of at least 15 different types of serotonin receptors. One drug, methysergide, works by blocking the activity of one kind of serotonin receptor while mimicking the effect of serotonin at a completely different receptor type. This last effect relieves or lessens headache pain by constricting the throbbing blood vessels that are responsible for the migraine attack.

The discovery of so many different serotonin receptors spurred researchers to develop a substitute for methysergide that would activate only the "bad" kind and leave the other receptors alone. Thanks to that effort, treatments for migraine are no longer haphazard and unreliable. At long last, migraine can be treated with specific drugs that work by altering specific neurotransmitter-receptor interactions. Sumatriptan, the first of a line of triptane drugs specific for migraine, has been followed by Zomig, Amerge, Maxalt, and others.

All these medicines work by targeting serotonin receptors on the trigeminal nerve, which mediates sensation from the head and face. During a migraine attack, this nerve is hyperactive, releasing protein fragments called peptides, which further increase the swelling of blood vessels and thereby worsen the migraine pain. Unfortunately, all presently available triptanes also target serotonin receptors on other

■ FOLLOWING PAGES: Commercial fisherman Steve Shears treats his seasonal affective disorder—a form of mental depression related to winter's shortened days— with therapeutic goggles that bombard his brain with healing light.

blood vessels as well, inducing side effects that can range from flushing and chest pains to, rarely, heart attacks and strokes. In the near future, researchers hope to come up with a triptane-like drug that is truly specific, affecting only the serotonin receptors of the trigeminal nerve and thereby being free of significant side effects.

Already, our understanding of migraine has grown to the point that doctors can pinpoint a specific serotonin receptor and chemically manipulate it, to relieve dreadful and incapacitating pain. Each advance in our understanding and treatment of migraine helps patients achieve control over an illness that can wreak havoc on their personal and professional lives. Best of all, that relief comes with none of the dangers and mumbo jumbo that plagued migraine treatments in ages past. Indeed, our current success in treating migraine should serve as a model for the treatment of depression, psychosis, and other debilitating diseases. What's needed for all these is a very specific drug that affects only the dysfunctional receptors, no others.

The closest thing we have to that are the breakthrough drugs used to treat Parkinson's disease. Parkinson's results from a loss of specific, dopamine-producing neurons in the substantia nigra, an area in the midbrain region of the brain stem, which projects upwards to targets in the basal ganglia, specifically the caudate and putamen. Since basal ganglia are important in movement, this loss of dopamine exerts widespread and catastrophic effects, such as tremor, rigidity, and a general slowing down and reduction in movement. Too often, the Parkinson's patient becomes a caricature

of old age: A hunched and stiff, slow-moving figure with a weakened voice and tremors.

At the moment, neuroscientists aren't certain why the cells in the substantia nigra die off. One theory holds that people who are predisposed to Parkinson's have certain brain areas that are more vulnerable than others to accelerated aging or to the influence of toxins. Even normal aging is accompanied by an attrition of brain cells that occurs—according to this theory—from the generation of free radicals that damage cell membranes both in the brain and the rest of the body. In those at risk for Parkinson's disease, this free radical damage may focus on the substantia nigra.

Treatment of Parkinson's disease may not be perfect today, but it represents one of the triumphs of modern psychopharmacology. In the late 1950s, researchers in Sweden reported that levodopa, a naturally occurring precursor to dopamine, could actually reverse a Parkinson's-like state induced in animals. The animals were first given a drug that depleted their brains of dopamine and other neurotransmitters. As a consequence, they developed many of the symptoms of Parkinson's disease. But with the administration of levodopa, those symptoms disappeared.

Following this experiment, levodopa was tested on people who had Parkinson's disease. The results were equally dramatic and consistent: Their symptoms improved and, in some cases, even vanished. Today levodopa remains the most popular and effective treatment for Parkinson's. It is used rather than dopamine itself because, unlike dopamine, it can cross from the blood into the brain. Once it enters the brain, it becomes converted into the active neurotransmitter dopamine.

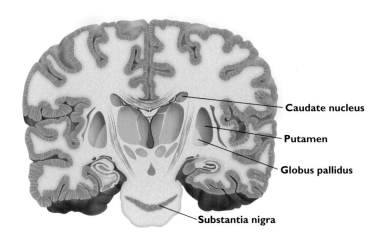

Caudate nucleus

Putamen

Globus pallidus

Substantia nigra

In the synapse, dopamine clings principally to D2 dopamine receptors. Normally, it is later released from the receptor and taken up by a nearby neuron, which breaks it down into its component parts by the action of two enzymes. Block even one of those enzymes, and some molecules of dopamine will remain whole; more dopamine will remain in the synapse, thus prolonging its action. This sequence forms the basis of action for one class of anti-Parkinson's drugs, which includes the brand names Eldepryl and Deprenyl.

Yet another and even newer type of drugs targets the second enzyme that normally breaks down dopamine, thus increasing the amount of available dopamine within the synapse. The result is an improvement in the symptoms of the disease. A third treatment involves agonistic drugs such as selegiline, which mimics the effect of dopamine at the dopamine receptor, and thus makes the body act as if more dopamine were present.

———————

Unfortunately, the magic of all these magic bullets is limited. After a few years or even just a few months, patients need more and more of the drugs to control their symptoms; the increased dosages mean that the side effects of these drugs—which can be serious—may become a problem. Strange writhing and twisting movements called dyskinesias may affect the limbs. Or the patient may have difficulty sitting still and will undergo a chorea—a series of widespread, quick, jerky movements that begin almost imperceptibly and progress to a twitching of various parts of the body. At its most extreme, chorea combines facial grimacing, unsteady walking, and abnormal posturing.

At other times the drug may cease to work altogether, allowing the Parkinsonian symptoms to return with a vengeance. The patient can be rendered completely immobile one moment and thrown into the abnormal movements of dyskinesia and chorea the next. Called the "off-on phenomena," this bizarre sequence of alternating signs results from various combinations of overstimulation, understimulation, and hypersensitivity of the patient's dopamine receptors.

In an attempt to improve the lot of the patient without developing drug-associated side effects, some neurologists have turned to surgery. One treatment, the fetal graft, transfers dopamine-rich tissue from a recently aborted human fetus into the patient's striatum, the receiving station of dopamine communication from the substantia nigra. Such operations have resulted in encouraging but modest improvement, and remain immersed in controversy due to the source of the grafts.

Yet another surgical approach uses electronic stimulators, implanting them in the subthalamic nucleus, a key component of the pathway linking the striatum (the major input structure of the basal ganglia) to the internal pallidum (the basal ganglia's major output structure). In April 1999 I witnessed the results of this technique.

Mahlon R. DeLong, a neurologist at Emory University's School of Medicine, showed me films of Sybyl, a middle-aged woman with severe Parkinson's disease, who had difficulty walking, talking, and feeding herself. But after stimulators were implanted in her subthalamic nucleus, Sybyl's tremors stopped. She moved quickly and smoothly, and could even run in the yard alongside her children. She became, in essence, a new Sybyl.

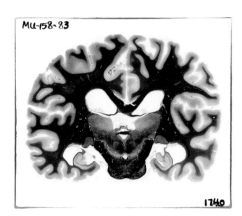
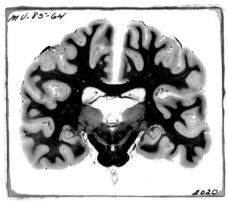

■ Physical scars of the devastation of Alzheimer's disease include numerous plaques and an obvious loss of brain volume, as seen in these brain slices from an Alzheimer's patient (right) and a healthy person (far right).

Surgical techniques for Parkinson's also include the severing of certain nerve fibers. While not reversible, this technique and the implantation of electronic stimulators are turning out to be "highly effective in reversing all of the cardinal features of Parkinsonism," according to DeLong.

Such successes in treating Parkinson's disease have led some neuroscientists to hope that a similar approach might work for another scourge of the mature brain, Alzheimer's disease. Since Alzheimer's is marked by a loss of the vital neurotransmitter acetylcholine, why not simply give the patient additional acetylcholine? Doctors tried, but the approach didn't work.

For one thing, Alzheimer's disease isn't restricted to a small part of the brain, as Parkinson's is. It involves many widely separated neurons and neuronal circuits, thus there isn't a single, defined site to receive a graft of acetylcholine-producing cells, for example. Nor are there discrete targets for surgical sectioning. Any cure for Alzheimer's is therefore likely to involve genetic and chemical rather than surgical approaches.

In addition, a medical or surgical approach to Alzheimer's is limited by the difficulty experienced in diagnosing it, especially in its earliest stages. Over 400 years ago, Montaigne noted what he termed the "obscurity" of dementing illnesses: "Sometimes it is the body that first surrenders to age, sometimes, too, it is the mind; and I have seen enough whose brains were enfeebled before their stomach and legs; and inasmuch as this is a malady hardly perceptible to the sufferer and obscure in its symptoms, it is all the more dangerous."

So we see that our understanding of the mind has made a quantum leap in the past two decades.

Prior to the discovery of neurotransmitters and their receptors, the brain—both in health and in sickness—was something of a black box. Although various drugs, plants, and home remedies exerted observable effects on thought and behavior (calming excited states, for example), no one had any idea of how they worked. All this changed when scientists discovered that drugs and other chemicals work by influencing specialized receptors within the brain. Practical applications of the emerging knowledge of neurotransmitter-receptor interactions include new drugs for brain diseases such as depression, psychosis, epilepsy, migraine, and Parkinson's disease.

Since all of these neuropsychiatric conditions represent neurotransmitter-receptor dysfunctions, our best chances for correcting them will come from the development of drugs that excite, inhibit, or otherwise influence neurotransmitters and their receptors. So far, neuroscientists have concentrated their efforts on only a handful of known neurotransmitters. It's likely, therefore, that as we learn more about other neurotransmitters, new insights will be forthcoming. One especially promising area is the subject of our next chapter: drug addiction.

■ Developed by Benjamin Rush, a founder of the American Psychiatric Association (and a signer of the Declaration of Independence), this 18th-century "tranquilizing chair" was used to restrain mental patients as they underwent copious bloodletting and purging. Other contemporary "treatments" included wrapping the mentally ill in cold packs.

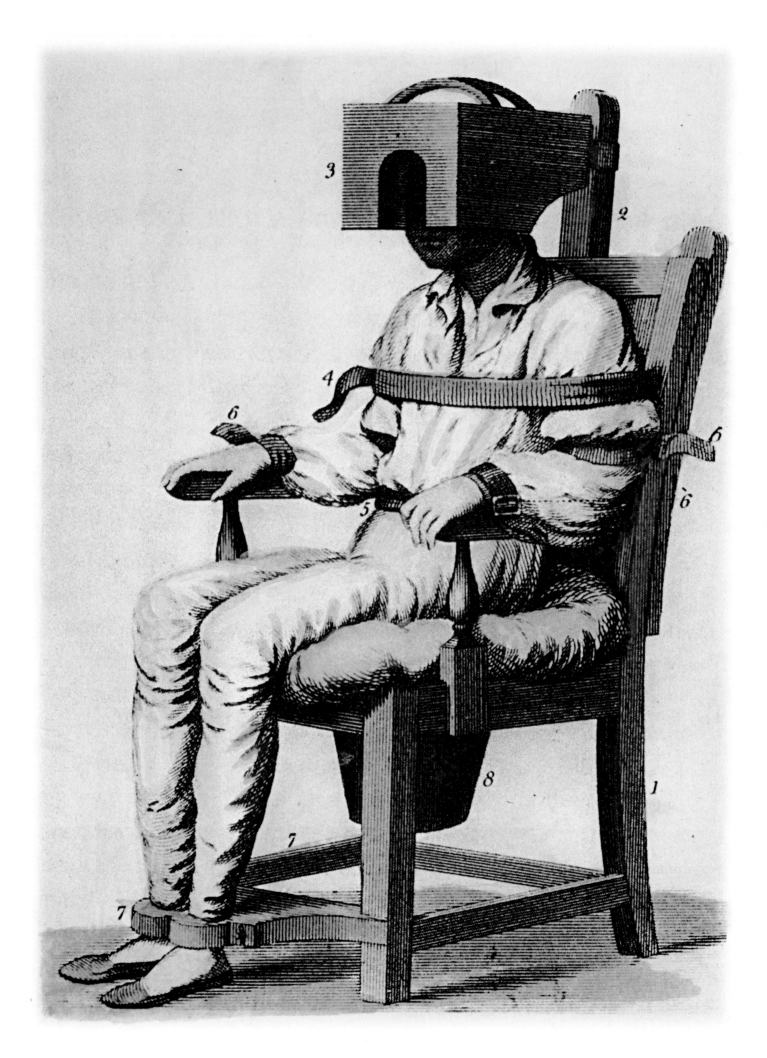

ADDICTION AND TREATMENT

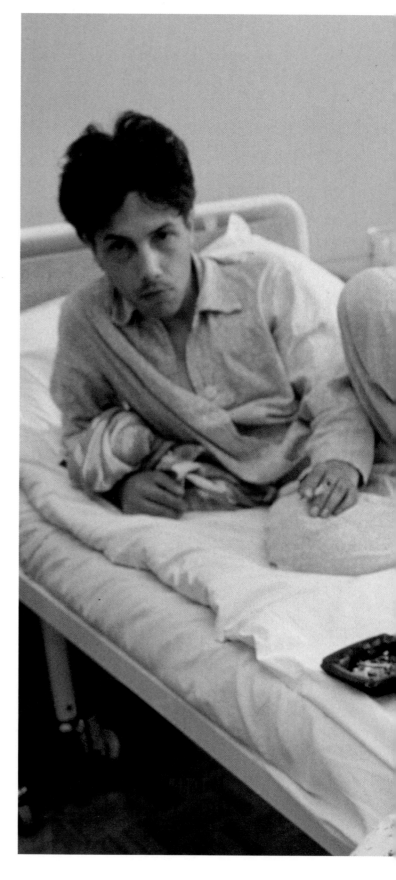

How many cups of coffee did it take to get you up and about this morning? Of course the chardonnay last night was excellent, but…five glasses? You say your headache this morning wouldn't go away after three aspirins but disappeared with a Bloody Mary and a few puffs on a Marlboro?

If one or more of these situations apply to you, you may have a substance abuse problem. Addiction or dependency isn't limited to illegal substances. In fact, most former morphine or heroin addicts will tell you that the hardest habit to break is that first morning drag on a cigarette. Caffeine and chocolate also can produce dependencies.

But here's the good news: Scientists recently have discovered the physiological basis for dependency. This means that treatments can be developed for our collective substance abuse problems. Simply put, substance abuse is a brain disorder.

"Virtually all drugs of abuse have common effects, either directly or indirectly, on a single pathway deep within the brain," writes Alan Leshner, of the National Institute on Drug Abuse. "The common brain effects of addicting substances suggest common brain mechanisms underlying all addictions."

All addictive drugs provide some temporary improvement in mood; in essence, a reward. Neuroscientists first suspected the existence of a reward system wired into the human brain as a result of research on animals. In the 1950s, American psychologist James Olds discovered that, once certain parts of the living brains of rats were implanted with electrodes, the animals would repeatedly press levers that delivered small electric currents to those areas. Key areas included the hypothalamus and other limbic system structures

Substituting one dependency for another? Although these detoxification patients in Warsaw, Poland, are receiving medical assistance for their heroin addictions, they enjoy unrestricted access to a substance even more widely addictive: nicotine.

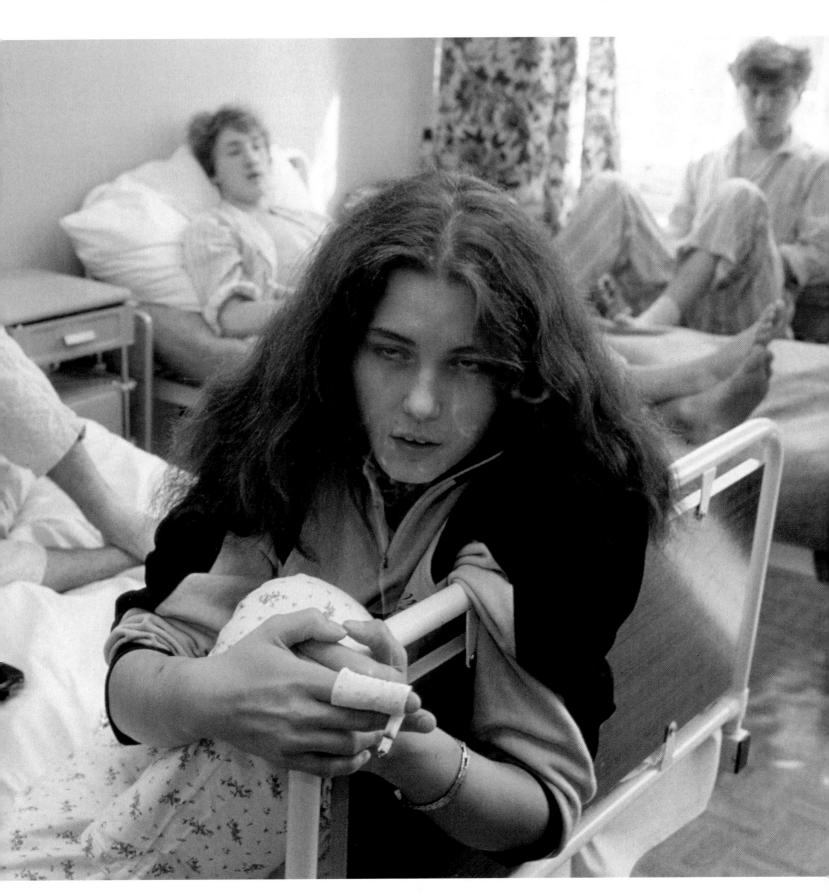

189

that are involved in naturally rewarding behaviors such as eating, drinking, and sexual activity.

Research on humans replicated Olds's findings. People undergoing brain operations, for instance, regularly reported pleasurable experiences in response to electrical stimulation of the same brain areas that had tested positive in animal research: the hypothalamus, the amygdala, and other sites within the limbic system.

Subsequent research led to an even more intriguing finding. Replace electrical stimulation with the injection of certain drugs into the brain, and the animal will press levers with equal enthusiasm. In addition, the same brain areas are involved. Further, the rate and amount of lever pressing to obtain a given drug serves as a measure of the intensity of the pleasure response produced by that drug. The more pleasurable the response, the more likely the animal will repeatedly press the lever—and, in the process, become addicted to the drug.

Known as the mesolimbic reward system, the brain circuitry that underlies the pleasure and addiction responses begins far down in the brain stem and extends up to a control center known as the nucleus accumbens. From there, impulses project upward and outward to the emotional brain center—the limbic system—then ascend to the most advanced region, the frontal cortex. Activation of this circuit is what underlies all types of addiction and substance abuse, whether the substances are perfectly legal, like alcohol, or are illegally acquired through underground channels.

In addition, activation of the mesolimbic reward system brings about significant changes in brain function, which persist long after the person stops taking the drug or other addicting substance. Many of us know smokers who have quit, in some instances for many years, and then returned to the habit after being exposed to the tiniest whiff of tobacco smoke perceived from across a room. Similarly, the practice by Alcoholics Anonymous to insist on a lifetime of sobriety for the alcoholic is based on the recognition that the risk of relapse continues lifelong. Relapse remains a possibility because both tobacco and alcohol modify the brain, specifically the brain's neurotransmitters.

Dopamine is one of the operative chemical messengers in the brain's reward system. When its transmission is facilitated, the addicting effects of cocaine, amphetamine, and nicotine are enhanced. Alcohol is a bit more complicated, for it involves other neurotransmitters as well as dopamine.

Whatever the abused substance, chronic use changes the reward system in specific and pernicious ways. Morphine, alcohol, and cocaine all induce changes in levels of proteins, neurotransmitters, and enzymes along the dopamine pathway. As a result, greater and greater amounts of the substance are needed to get the same effect. What's worse, if the person tries to stop, his craving for that drug increases even more.

These brain alterations in the reward system can bring about horrific feelings of depression, isolation, anxiety, irritability, and despair, all of which drive the abuser to what seems like the only possible response: a return to the substance of abuse. And so, the afflicted person holes up somewhere and drinks or drugs the afternoon away.

While dopamine is a key neurotransmitter in addiction, glutamate also plays an important role.

It's speculated that while bursts of dopamine attract the brain's attention and direct it towards obtaining drugs, alterations in the glutamate signaling system lead to compulsive drug use. It's likely that glutamate works by enhancing learning and memory so that the addict repeats the same ritualistic patterns of behavior associated with drug use. Memory for those rituals is then incorporated into the craving and subsequent addiction based on a conditioning response, whereby drug-related objects like needles and syringes elicit craving.

In one test of this conditioning response, addicts watched movies of people using drug-associated items like glass pipes and razor blades. PET scans of the addicts' brains revealed that the intensity of craving aroused by the movies paralleled the intensity of activity in the frontal cortex and the amygdala, two regions that release glutamate onto neurons in the nucleus accumbens. This suggested that glutamate-rich brain structures associated with thinking and memory play an important role in conditioned cravings elicited by reminders of past drug use. For some addicts, it may involve nothing more complicated than a walk through a neighborhood where drugs were formerly obtained; the neighborhood acts as a conditioning stimulus that elicits drug-seeking behavior.

Since glutamate plays such an important role, researchers are hoping that medications capable of interfering with glutamate signaling will also interfere with conditioning, and thus will block drug cravings. A successful medication would enable an addict to encounter drug paraphernalia or other conditioning stimuli without experiencing cravings.

An experimental drug known as MK-801 prevents rats and mice from becoming sensitized to cocaine and amphetamine, by preventing glutamate from acting at a key receptor. Another drug, which has been approved in Europe for treating alcoholism, works by dampening glutamate-triggered activity in cortical neurons. It's postulated that this mechanism contributes to the reduction of cravings. Interestingly, the same drug stimulates glutamate neurons in the hippocampus and nucleus accumbens, an action that is believed to convert neurons that are normally quiescent during withdrawal into an active and revitalized state, which helps fend off withdrawal symptoms and cravings.

But because people differ in their susceptibility to substance abuse, not everyone who tries a drug gets hooked. People also differ in their ability to stay off an abused substance. Some walk away from it without relapsing; others never achieve complete recovery. Why is this so?

Neuroscientists believe it's because substance abuse leads to changes in gene expression at the molecular level. Since no two people, not even identical twins, possess the same genetic makeup, it should come as no surprise that the genetic changes wrought by drugs differ from person to person.

Genetic contributions to substance abuse are also suggested by studies of alcoholism in adoptees. In one famous study, 62 percent of adopted males whose biological parents were alcoholic turned out to be alcoholic themselves. This held true even when the adopted parents were teetotalers. In contrast, only 24 percent of the adoptees whose biological parents were not alcoholic became alcoholic.

Ken Blum, a pharmacologist formerly at the University of Texas at San Antonio, first linked

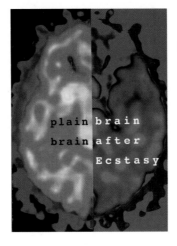

Part of one campaign in the ongoing war on drugs, these scans dramatize the ravages of Ecstasy, a synthetic amphetamine analog that creates a "rush" by overstimulating serotonin-producing cells. The left half shows active serotonin sites in a healthy brain; dark areas on the right are Ecstasy-damaged sites, lingering even after three weeks without the drug.

alcoholism to a specific gene several years ago. Termed the A1 allele, it is one of four naturally occurring forms of a gene that encodes for a particular dopamine receptor. Possessing this gene doesn't consign you to a lifetime of alcoholism, but it does make you more vulnerable to alcohol addiction. Blum has suggested that the A1 allele is a defective "reward gene" that may underlie a "reward deficiency syndrome" that includes many different forms of substance abuse and compulsive disease. Research teams in France, Japan, and Finland have confirmed the linkage of the A1 allele to substance abuse.

How do genes exert such a strong determining influence? One possible mechanism is that long-term substance use induces changes in gene expression, which causes more or less of a certain neurotransmitter to appear in key brain areas. Alternatively, the change in gene expression may alter the structure and number of synaptic connections formed by individual neurons. In either case—and both mechanisms may be operative—the elements emerge for a battle plan against substance abuse.

Pharmaceutical firms are now closer than ever to attacking dependency at the level of specific brain targets and neurotransmitters. At the moment, several intensely competitive efforts are underway aimed at alcohol, cocaine, opium-like compounds, and—the most widely abused addictive substance of all—nicotine.

If you're a smoker and have tried unsuccessfully to quit, you've experienced firsthand the powerful effect a neurotransmitter can have on its receptor. And you're not alone. Of the 45 million smokers in the United States, about a million stop each year.

But from this pool of "former" smokers, about 2,000 relapse each day. In addition, 3,000 people a day take up smoking, almost all of them adolescents. Most new smokers begin between 14 and 16 years of age. Sadly, the number of new smokers is almost matched by the number of people who die daily from smoking-related causes.

If you could ask only one question of a smoker in order to determine his chances of successfully quitting, what question would that be? Surprisingly it's not "How many cigarettes do you smoke a day?" but "How soon after waking in the morning do you smoke that first cigarette?" Among those who keep cigarettes on the night table and light up within a few minutes of awakening, the chances of successfully breaking the nicotine habit without help is almost nonexistent. That's because their brain's craving for nicotine is so strong that even a brief delay causes intense discomfort, strongly compelling them towards that first light-up of the day.

Nicotine is as addictive as cocaine or heroin, according to David Smith, the founder of the Haight-Ashbury Free Clinics, a community-based health care and addiction treatment facility in San Francisco. One of the world's experts on addiction, Smith came up with the "Three C's" concept of addiction: Compulsion, loss of Control, and Continued use despite adverse consequences.

Alexander Glassman, a Columbia University specialist in nicotine addiction, agrees that addiction of any type involves "compulsive, uncontrollable drug craving, seeking, or using despite adverse health or social consequences and often both." Smith and Glassman concur that drug use should not be considered addiction unless the user is fully aware of the need to stop the drug use but can't.

Glassman adds that addiction is also characterized by its chronic relapsing nature. "Even after months or years of abstinence, single re-exposure can produce an almost immediate re-addiction."

On the molecular level, nicotine addiction involves the coupling of nicotine with one of two specialized types of acetylcholine receptors within the brain. In fact, scientists classify acetylcholine receptors on the basis of their sensitivity to nicotine and muscarine, a toxin derived from mushrooms.

––––––––––––

Nicotine is also a poison. A component of many pesticides, it retains its poisonous potential even in low dosages. For centuries, gardeners have used cigarette ashes to protect plants from nibbling by animals and insects. Certainly most first-time smokers can attest to nicotine's toxic effects, which in the neophyte smoker can include severe nausea, vomiting, headache, and tremor. But the body usually becomes accustomed to the poisonous effects of low levels of nicotine.

If one's nicotine intake exceeds certain limits, however, it can be deadly. If a young child swallows the contents of its parent's cigarette pack, the drug rapidly achieves high concentrations in both the blood and brain. Hallucinations, respiratory stimulation followed by respiratory arrest, convulsions, severe slowing of the heart, and even death can result.

Given that we know so much about nicotine's ill effects, you might expect that people would avoid it like the plague. But nicotine, in common with all addictive substances, produces euphoria. Smoking delivers a rush of nicotine to receptors, altering them and increasing their number. Since the receptors are located in the pleasure mediating center of the brain—the mesolimbic system—smoking induces a minor "rush" followed by a slow decline that lasts until the nicotinic receptors turn back on in anticipation of the next delivery of nicotine. After many years of smoking, the number of nicotine receptors in the smoker's brain increases, as does the number of cigarettes smoked. This increase in nicotine receptors helps explain why breaking the habit becomes so difficult.

When a smoker stops smoking, his receptors essentially cry out for their usual nicotine hits. The molecular processes associated with addiction exert their effects several levels higher, in the behavioral sphere, and the smoker experiences anxiety, edginess, irritability, depression, and other symptoms of withdrawal. He—and his receptors—desire more of the addicting substance.

Quitting smoking isn't impossible; if a smoker persists in not smoking, the nicotine receptors in the brain's pleasure centers will eventually decrease through a process called down regulation. Simply put, the number of receptors wanes in the absence of a steady infusion of nicotine. On the subjective level, this decrease is experienced as a lessening and eventual disappearance of the craving.

■ FOLLOWING PAGES: Best man brings the brandy—planted in the wedding cake— to toast bride and groom at a wedding in the Georgian Republic. Among the most available and popular means of altering the mind, alcohol taken in moderation can encourage relaxation and conviviality. In excessive amounts, however, it enslaves the mind while poisoning the brain.

> **The more cigarettes you smoke, the more likely you are to have a history of depression.**
> —Alexander Glassman

In fact, the number of nicotine receptors in a smoker who has successfully given up cigarettes will decrease nearly to a level found in the brains of non-smokers. It will never be identical to that of a person who never smoked, for even a brief exposure to nicotine—perhaps no more than a few cigarettes—will prod a person's genes into escalating the number and rapacity of nicotine receptors. This rapid and unforgiving response helps explain the high failure rate among smokers who try to quit. Addiction to nicotine can be even harder to overcome than other addictions because it is legal, relatively cheap, and—until recently—tolerated by non-smokers.

In short, nicotine possesses all the features of an addicting drug. It induces euphoria, physical dependence, and withdrawal effects. Like drug addicts, smokers are not deterred by the knowledge that their habit can induce various illnesses.

In addition, nicotine produces varied and sometimes contradictory effects, based on the many different nicotine receptors in the brain. It exerts a calming effect, yet it speeds up the heart rate. It can enhance concentration and performance, but it also induces tremors and interferes with sleep. Its actions depend, quite simply, on the location and type of receptor it links up with.

Nicotine's most powerful effect takes place at the nicotine receptors in the brain's pleasure centers, principally the nucleus accumbens. Recall that cocaine, amphetamine, and other drugs of abuse act here too. Indeed, nicotine given intravenously has been compared to low doses of cocaine. Interestingly, intravenous nicotine is pleasurable to regular smokers but not to non-smokers.

Nicotine's potential for addiction varies from smoker to smoker, ranging from one cigarette every ten minutes—the chain-smoker—to one every six to eight hours. Even the interval between puffs can vary. If low-nicotine cigarettes are chosen, both the number of cigarettes smoked and the frequency of puffing are likely to increase, because the smoker seeks to achieve the same nicotine level in his brain.

Whatever the nicotine content of a particular cigarette, the nicotine in tobacco smoke inhaled into the lungs rapidly makes its way into the body's fat-rich tissues, especially the brain. This rapid distribution and uptake by the nicotinic receptors, when coupled with nicotine's rapid breakdown, accounts for the tendency of smokers to quickly develop a dependency.

In addition, there's evidence that a particular gene may predispose some people to smoking. This gene codes for DRD2, one of the five known dopamine receptors. In one study, smokers were found to be much more likely to possess a variant of the DRD2 gene. Further, those with a certain form of that gene had begun smoking at an earlier age and were much less likely to successfully quit.

A similar DRD2 variant can be observed in other forms of addiction, including alcoholism, compulsive overeating, and pathological gambling. Not every person with these problems has a variant DRD2 gene, however, nor is this the only factor that enhances susceptibility to addiction. Genetics is thought to account for only about half of a person's susceptibility to nicotine addiction. The other half arises from environmental variables, such as imitating the smoking behavior of celebrities, or other encouragement to smoke in early adolescence. Not all of the genetic research points to an enhanced susceptibility to nicotine; some suggests that certain people may be *less* likely to smoke, thanks to their genetic endowment. Researchers at Georgetown University found that people with a variation of the gene that transports dopamine are less likely to start smoking and, on average, 1.5 times more likely to quit if they do start, compared to people who lack this gene variant.

Such findings are consistent with the theory that dopamine is critical both in starting and in stopping the smoking habit. Further, the discovery of genes for addiction opens up the possibility for screening programs aimed at identifying the people at highest risk for nicotine addiction. After testing, those with the standard genetic profile would be treated conventionally; those found to possess a genetic variant that places them at higher risk would be targeted for special treatments.

———————

Here's another way to think about the brain of a smoker. Each time a person smokes, he or she reinforces functional connections within a pathway originally established at the moment of that first drag on that first cigarette. This scenario is certainly in line with a basic principle of brain operation: At every moment of our lives, the brain is establishing cell networks composed of millions of neurons. Each action, each thought, each memory involves the activation of vast networks of cells. And each time we act or think or remember, we strengthen certain cell connections somewhere within our brains. This explains how habits are formed and why they are so hard to break. Only when the cell pathway disappears or is significantly weakened can the habit be broken.

And until the nicotine habit is broken, the heavy smoker trying to quit can expect a lot of misery. When deprived of regular hits of nicotine, various kinds and degrees of withdrawal occur: Decreased blood pressure and pulse, decreased alertness and reaction time, weight gain, diarrhea, headache, and nausea. In addition to these physical changes, a nicotine-deprived smoker sleeps poorly and often becomes hostile, aggressive, and anxious.

———————

There is yet another complication to nicotine addiction: Nicotine is an antidepressant, and so its use can mask underlying depression. In the words of nicotine addiction specialist Alexander Glassman, "The more cigarettes you smoke, the more likely you are to have a history of depression."

Among Americans, about 45 percent of the women and 64 percent of the men are or have been smokers, while more than 70 percent of the women and 80 percent of the men have a history of depression. What this means is that depressed people may be using nicotine for its antidepressant effect; when they wean themselves off cigarettes, the depression manifests itself. This is one reason why I—even as a nonsmoker—think we may be moving a bit too fast in all our anti-smoking efforts. Depressed people who, in addition, are addicted to nicotine, feel badly enough about themselves without having to suffer the additional burden of pariah status. There is a need for compassion.

While studies of twins reveal a genetic link between depression and smoking, they aren't very helpful in determining why depressed people experience such great difficulty stopping smoking. One thing is certain: Depression is not a withdrawal symptom. Smokers who normally don't suffer from depression don't develop it when they stop smoking. They may feel edgy, irritable, or temporarily "down," but they don't develop clinical depression. Discrepancies between depressed and non-depressed people who try to kick smoking probably reflect individual susceptibilities to drugs like nicotine.

Faced with the number and variety of unpleasant physical and emotional effects that accompany attempts at quitting, many smokers simply give up and head to the corner store for another pack—a relapse rate estimated nationally at about 2,000 people per day. But alternatives do exist.

Most treatments for smoking addiction involve the administration of nicotine in some form other than cigarettes. This is acceptable because it's not the nicotine in cigarettes that leads to cancer and emphysema, but the tars and other toxic chemicals released by the burning of tobacco. These toxins irritate the respiratory tract during a process that, after years of exposure, leads to cancer. Toxins are also largely responsible for the increased incidence of heart attacks and strokes among heavy smokers. Nicotine alone—separated from its toxic accompaniments found in cigarette tobacco—is addictive, but it isn't thought to be responsible for smoking's major health consequences.

That's why nicotine gum can be an effective treatment—as long as the smoker chews frequently enough to achieve the necessary blood level of nicotine. Heavy smokers need more gum and have to chew more actively; 20 to 25 pieces a day, each chewed for 30 minutes, isn't unusual. Not surprisingly, most people experience jaw discomfort from all that chewing, especially since nicotine gum tends to be harder than regular gum. They can also experience the loss of dental fillings and even arthritic changes in the jaw. And if any gum is accidentally swallowed, they can suffer nausea, vomiting, and a resultant disinclination to try any more gum.

The main limitation on the effectiveness of nicotine gum is the mouth's need to remain at an alkaline pH. Since the gum may not deliver enough nicotine to maintain a smoker's usual level of alertness, he or she may counter grogginess by reaching for a cup of coffee. Coffee, colas, and citrus drinks are acidic; taking them interferes with the absorption of nicotine from the gum, resulting in the reappearance of nicotine-withdrawal symptoms.

Use of a nicotine patch, which dispenses nicotine through the skin, doubles the success rate compared to a placebo, and can do even better if the smoker also takes part in a behavioral treatment program. Over successive sessions, the smoker works with a therapist on ways of managing situations that in the past have led to a relapse.

Another option is nicotine nasal spray. One of the most successful treatments, however, involves a combination of the patch and the antidepressant drug bupropion (Wellbutrin), marketed as Zyban. It is thought to work by keeping dopamine levels elevated in the brain, thus emulating the physiological reward state induced by nicotine. When using both Zyban and the patch, depressed as well as nondepressed smokers quit at an astounding rate— 58 percent remained abstinent at ten weeks, the time when most relapses occur.

In summary, researchers generally agree that nicotine, though addictive, is less harmful in low doses than cigarette tars, which contain dozens of

Even after months or years of abstinence, single re-exposure can produce an almost immediate re-addiction.
—Alexander Glassman

cancer-causing agents. Second, recent research on Nicorette gum and Nicoderm patches indicates that many smokers use them to temporarily combat nicotine cravings in "No Smoking" situations, rather than as tools for complete smoking cessation. Terms like "exposure reduction" rather than total smoking cessation may be a more achievable goal, according to some pharmaceutical executives.

"If it's the nicotine people want, why not give it to them?" asks a former director and scientific information officer for Pharmacia & Upjohn, quoted in a *Wall Street Journal* article. "I think we need to take care of people who've become so addicted to nicotine that there's nothing you can do but give them a maintenance drug," comments a former smoker.

———

Giving people what they need and want underlies the argument for legalizing drugs such as marijuana, cocaine, and heroin. But this argument—often buttressed by statistics from the Netherlands, where **cannabis** has been legal since 1976—is less compelling when put into the context of recent brain research. George F. Koob, an addiction researcher at the Scripps Research Institute, adds, "The more we discover about the neurobiology of addiction, the more common elements we've seen between THC [the active ingredient in **cannabis**] and other drugs of abuse." Included among those elements is marijuana's ability to produce brain changes that could lead to withdrawal reactions.

According to recent experiments, **cannabis** elicits the same dopamine reaction as nicotine, cocaine, and heroin: a flood of dopamine into the nucleus accumbens of the midbrain. Marijuana withdrawal also elevates reward thresholds and acti-

vates stress systems in the brain. These changes could account for the three C's of addiction—Compulsive use, loss of Control, and Continued use—coined by David Smith.

But while debate continues about whether or not substances like marijuana are highly addictive, there is no debate among brain researchers about their ultimate research goal: a detailed understanding of the molecular and cellular mechanisms of substance abuse. If vulnerable brain cells and brain circuits can be identified, the potential is there for treatments that can prevent or reverse the brain changes caused by addiction.

It's likely that, within the next decade, brain scientists will come up with an all-encompassing, neurologically based understanding of substance abuse. From there it should be reasonably easy to fashion chemical antidotes. That's because, as a result of recent research, scientists now know the specific networks involved in substance abuse.

In the January 1998 *American Journal of Psychiatry*, scientists from Harvard described a safe and effective brain-imaging technique that highlights active areas by sensing differences in local blood flow and hemoglobin levels. Testing cocaine addicts, they found increased brain activation in parts of the frontal lobes and the anterior cingulate. Even more intriguing, these areas "lit up" whenever the addicts experienced cocaine-related sights and sounds. In essence, the imaging gives us color-coded snapshots of those brain areas that are responsible for craving.

Cocaine—along with opiates, alcohol, nicotine, and barbiturates—all reduce metabolic activity of the brain by about 10 to 15 percent. Moreover, this metabolic downturn correlates directly with the intensity of the drug high.

No one knows for certain why this happens. One theory is that stimulating the brain's dopamine pathways activates the deep-lying pleasure pathways while toning down the influence of the cerebral cortex, where conflicts and worries are elaborated. Whatever the reason, neuroscientists have already reached the point where substance abuse can be detected by tests that are safe and easy to administer. And, after more than a decade of research, they are much closer to achieving a primary goal: synthesizing a drug that will reduce drug craving.

Until recently, all approved drugs that tried to reduce drug craving were replacement therapies, that is, drugs that were like the addictive drug except that the withdrawal process for them either doesn't take place or occurs in a milder form. Methadone, for example, works as a substitute for heroin; after successful substitution, the addict is continued on methadone for a time and then later withdrawn. Methadone withdrawal is quicker, safer, and easier than withdrawal from heroin.

This is different from the techniques used to help smokers quit, for nicotine patches and other treatments deliver the addictive drug itself—not a substitute—in gradually decreasing amounts, in order to ease withdrawal. Current research on nicotine addiction is looking for synthetic drugs that act only on specific subtypes of nicotinic receptors and are not addictive.

Incredible progress has occurred in regard to another highly popular addicting substance, alcohol. Starting in the 1980s, researchers began looking for genetic patterns to alcoholism. They discovered that, among adopted males, a substantial majority—

■ Aumara Indians in Bolivia socially chew coca leaves—source of cocaine—as they have for centuries. Ingesting drugs to expand the mind is a tradition thousands of years old, one that has affected nearly all civilizations, at every level of human society.

62 percent—developed alcoholism if either of their biological parents had been alcoholic. Predicting which children are at greatest risk for developing alcoholism isn't easy, but researchers have assembled some rough guidelines. According to Robert Cloninger, a psychiatrist at Washington University, in St. Louis, Missouri, alcoholism can take two distinctly different forms.

Type 2 alcoholics are typically reckless, self-destructive, and antisocial. They start drinking at an early age and lose control over their drinking soon thereafter. On psychological tests, Type 2 alcoholics tend to score high on "novelty seeking" and score low on "harm avoidance."

Type 1 alcoholics take a more cautious, anxious approach to life. Rather than adopting the Type 2's "hell with you" approach, they try to get on with others and generally seek social approval.

Cloninger believes it is possible to distinguish these two temperaments in children as young as ten years of age. If true, such identification could provide the basis for early intervention aimed at preventing the transition from a predisposing personality pattern to the establishment of overt alcoholism. Intervention would be especially crucial in instances where this predisposition is combined with a genetic background of alcoholic parents. This would be most applicable to Type 2 alcoholics, who are more likely than Type 1 to have alcoholic parents.

Another approach to alcoholism tries to understand it on the chemical and molecular levels. The brain's processing of alcohol undoubtedly involves several neurotransmitters in addition to dopamine. We know this because laboratory rats, given a chance, will self-administer alcohol even after destruction of part of their dopamine pathways.

In addition to affecting dopamine levels, high doses of alcohol increase the affinity between GABA and its receptors. Typically, this leads to an opening of the chloride channel in the cell membrane—an action that increases neuronal inhibition. As alcohol intake increases, the channel becomes active even in the absence of GABA. Additional mischief occurs when alcohol is mixed with drugs such as benzodiazepines, because they happen to share the same GABA receptor; just as alcohol increases that receptor's affinity for GABA, it increases its affinity for Valium-like drugs. Taking alcohol with Valium can lead to a potentially fatal depression of one's normal breathing rate. That's just one reason why benzodiazepines should never be taken with alcohol.

Another consequence of sharing the same receptor complex is that alcohol and benzodiazepines exhibit cross-tolerance: Chronic exposure to either drug results in tolerance to the acute effects of the other. Thus, a person drinking alcohol needs a higher dose of Valium in order to achieve a given degree of tranquilization. This can cause overdose.

On the positive side, cross-tolerance means that the symptoms of alcohol withdrawal can be lessened by cautiously and gradually substituting a benzodiazepine for the alcohol. In fact, it's the basis of modern methods of detoxification. Once substitution is made, the patient is slowly weaned from the benzodiazepine under medical supervision. After completion of this "drying out," the alcoholic—now alcohol- and drug-free—is ready to begin a sustained rehab program.

Part of that program is to convince the alcoholic that he remains permanently at risk for future addiction. In most instances this means no alcohol for the rest of his life. In a few instances—and this is controversial—some alcoholics resume controlled drinking without sliding down the "slippery slope" into renewed alcohol addiction. But for the vast majority of people recovering from serious alcohol problems, the prospect of an occasional drink now and again just isn't worth the unknown risk for future addiction.

For alcoholics who need a drug to help keep them away from booze there's naltrexone, approved in 1984 for the treatment of opiate addiction. It also reduces alcohol craving, although how it does so isn't completely understood. Now marketed by DuPont as ReVia, it is the most effective drug yet found for alcoholism. When combined with support groups and psychotherapy, it offers the prospect of a successful brain-based treatment for alcoholism. Until the development of naltrexone, in fact, only one other medicine was available to treat alcoholism. That drug, Antabuse, worked by inducing vomiting in anyone imbibing an alcoholic beverage.

Will other drugs soon be available to treat alcoholism without undesirable side effects? While a certain degree of confidence about that prospect seems justified, overconfidence is inappropriate. Alcoholism is a form of drug addiction and, in the words of researcher George F. Koob, drug addiction "is a chronic relapsing disorder characterized by compulsive uncontrollable use of a drug."

Relapses occur because all addicting drugs activate brain systems that are involved in processing rewards. Whenever the reward system is chemically activated, formerly rewarding experiences lose their appeal. In essence, once drugs have turned on your reward systems, why would you bother to do the things that normally activate those systems?

For instance, when you listen to a symphony or attend a baseball game, parts of your brain—notably the amygdala and the ventral tegmentum—become activated. But if you're on drugs that have already turned on your reward system, there's no need to bother with music or a sports event. What's more, the drug experience establishes a conditioned reflex in the brain, so that anything that reminds you of a previous drug-use episode creates and strengthens a craving for that drug.

Taken over time, alcohol, morphine, and cocaine all induce alterations in proteins and enzymes within the pathways for dopamine and other neurotransmitters. These alterations in the brain's chemical structure and functioning underlie the addict's "memory" for his addicting substance. Such changes also explain why substance abusers invariably develop a tolerance to their drugs and have to take more in order to recover the euphoria and good feelings associated with early use. In essence, the brain of the addict or alcoholic has been chemically restructured and a new, pernicious form of memory established.

"We should be thinking of addiction as a kind of memory, a disordered, maladaptive memory that gets us into trouble. But it's a memory that is, at least in part, involuntary," writes addiction specialist Charles P. O'Brien, of the University of Pennsylvania.

If O'Brien is correct, it's obvious that drug memories are responsible for some of the most serious problems facing our nation today: Gang wars to control drug sales and turf; billions of dollars in lost revenue, health care, and law enforcement expenses; community disruption; the loss of minds and brains to deadly overdoses or to the hell of continued addiction. All this is the result of individual attempts to achieve transient alterations in consciousness. With so many negative aspects to drugs, a simple question springs to mind: Wouldn't we all be better off if the human brain wasn't so susceptible to addiction? Surprisingly, the answer to that question is a resounding no.

Certainly our propensity to abuse certain substances is an unfortunate by-product of our brain's design. But that design is perfect for achieving the adaptive behavior most likely to lead to reproductive success. Dopamine, for instance, is associated not only with the pleasures of eating but also with the pleasures of sex, thus it indirectly prods us into courting and mating and reproducing. Indeed, dopamine is involved in many if not all of our most rewarding and pleasurable activities. Problems arise when our brain encounters drugs capable of hijacking those pleasure circuits that are involved in the normally adaptive joys of eating and sex.

"These drugs are not intrinsically evil," believes Jeff Victoroff, director of neurobehavior at Rancho Los Amigos National Rehabilitation Center, in Downey, California. "They are merely the distilled essential liquors of reward which, arguably, we have a tremendous motivation to find and exploit in our environment."

In a nutshell, drugs of abuse set off false signals within the brain's pleasure centers. This leads to the replacement of normal, adaptive behaviors by drug-seeking ones. Unfortunately, drugs aren't alone;

other aspects of our contemporary environment can lead to similar difficulties.

More and more people today readily talk of being "addicted" to everything from video games to shopping to unhealthy fast-foods. University of Michigan psychiatrists Randolph Nesse and Kent Berridge write in their classic paper on the use of psychoactive drugs, "We are vulnerable to such fitness-decreasing incentives because our brains are not designed to cope with ready access to pure drugs, video games and snack foods."

In light of the powerful combination of the biological and social forces involved in addiction, is it realistic to believe a "war" can be won against drugs? In answering that question, remember that drug use is rooted in the fundamental design of the human brain. That brain is designed to maximize pleasure and minimize pain. Therefore, any substance, natural or man-made, that plugs into the brain's pleasure centers can prove a formidable opponent. Rather than seeking victory in a war on drugs, perhaps we should concentrate on new ways to develop sensible strategies aimed at preventing and treating addiction.

■ Getting his mind off work, a Washington, D.C., businessman snorts cocaine into his nose, where sensitive membranes absorb the drug and pass it into his bloodstream. Because mind-altering drugs influence the brain's pleasure centers deep beneath the cerebral hemispheres—where reason and language are mediated—psychotherapy alone rarely succeeds in getting abusers to kick their hard-core habits.

FIXING THE BRAIN

The brain is the most daunting bodily organ. Hidden beneath a thick skull and three protective membranes, it resists any casual attempts to plumb its mysteries. As a result, few people have personally seen or touched a living human brain. The chief exception is the neurosurgeon who encounters the brain on a daily basis. But in order to exercise that privilege, a neurosurgeon must undergo more than a decade of mentally and physically demanding training.

Compared to other surgical disciplines, neurosurgery has a short history; most of its advances have taken place in the past 70 years. But the earliest "brain operation"—the drilling of holes into the head, or trephination—was practiced by many ancient civilizations, including early European, Egyptian, Indian, and various South American Indian cultures. Artifacts establish the extensive use of this crude technique for relieving cranial pressures in pre-Incan civilizations as early as 2000 B.C. Surgical instruments were made of bronze or obsidian—a hard, sharp-edged volcanic rock.

The typical patient, a soldier who had suffered a head injury in battle, might experience bouts of drowsiness, lose strength on one side of his body, or develop other signs of increasing pressure within his

■ During a radical, day-long procedure, neurosurgeons prepare to remove one of the cerebral hemispheres of their young patient, an extreme epileptic whose seizures have not responded to anti-epileptic drugs. Brain surgery, performed rarely only a century ago, now has become routine.

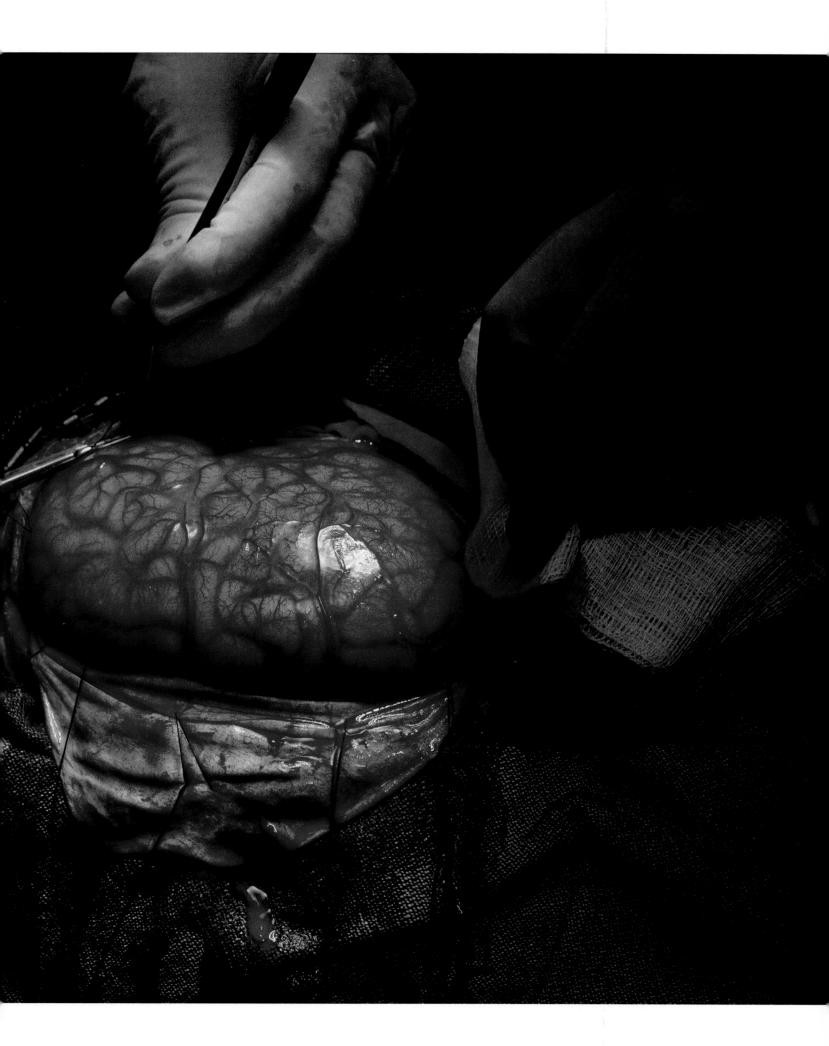

Moor him at his mooring stakes until the period of his injury passes by. Thou shouldst treat it afterward with grease, honey and lint every day, until he recovers.

—Edwin Smith papyrus

skull. He would then be restrained and given alcohol or some other preparation aimed at inducing relaxation, sedation, and—if things went really well—perhaps even sleep.

Next the surgeon, using the sharpened trephine, scraped or bored a hole into the skull. The job had to be done carefully; a rough or impatient thrust could do more harm than good, plunging the sharp trephine not just through the skull but also through the outermost of the brain's coverings, the dura mater. Infection and death usually followed, as the pierced dura left the brain exposed to bacteria and other organisms.

Although crude by today's standards, these ancient forerunners of modern neurosurgical operations did not always kill. We know this because archaeological discoveries of ancient skulls show healed trephine openings—proving that these early patients recovered from their operations and went on to live for many years.

We also know that the procedure was common. Of 120 skulls found in a single dolmen, or ancient tomb site, in France, 40 contained trephination holes. The vast majority of the skulls were normal, except for the holes. So why were the trephinations performed? Originally, such primitive operations may have been aimed at releasing "evil spirits" or "bad humors" from the head. Religious and magical explanations of many civilizations assessed epilepsy and other disorders as due to demons within the skull. What more logical treatment than to drill holes in the skull and thereby let the demons out?

Another explanation holds that trephination was used to alleviate tension headaches common in primitive times. Perhaps the ancients believed that the creation of holes in the skull served to release

pressure within the brain. And judging by contemporary experiences with placebos, it's probable that at least 60 percent of their patients reported relief!

Whatever the precise motivation, it's likely that these Neolithic neurosurgeons trephined the skulls of people who complained of headaches, suffered seizures, or experienced visual or auditory hallucinations. We can assume that, in some cases, cures or temporary relief could result from trephinations. For instance, a trephination for a headache that was actually due to a brain tumor would relieve the increased pressure on the brain and thereby provide some relief for the patient.

———————————

Some 3,000 years after that French dolmen was built, the ancient Sumerians and Egyptians achieved a working understanding of the brain and elementary neurosurgical principles. In fact the Edwin Smith papyrus, a collection of medical case histories composed around 3500 B.C., contains the first recorded use of the word *brain*. Most of its descriptions involve battlefield injuries, including 13 cases of skull fracture. Each case includes a description of the injury and the wound, a diagnosis, prognosis, and treatment. Here is one example:

"If thou examinest a man having a gaping wound in his head, penetrating to the bone and perforating his skull, thou shouldst palpate his wound; shouldst thou find him unable to look at his two shoulders and his breast and suffering with stiffness in his back....

"Thou shouldst say regarding him: 'One having a gaping wound in his head, penetrating to the bone and perforating his skull, while he suffers with stiffness in his neck. An ailment which I shall treat.'

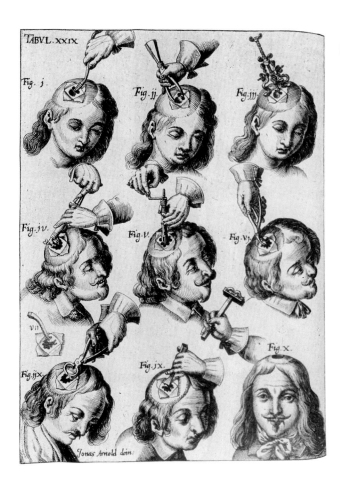

patients relaxed into an alcoholic stupor, others became more agitated, and difficult to control.

Modern neurosurgery depended on three medical advances. First came the use of bacteria-killing solutions, spearheaded by Lord Joseph Lister (who would be immortalized two centuries later in Listerine). Second, sterile surgical technique reduced the likelihood of infection. Finally, anesthesia, developed by William T. G. Morton, led to more dependable painkillers and more reliable sleep-inducing agents that enabled surgeons to undertake longer and more complex operations.

———————————

Of the three advances, anesthesia played the most direct role in the development of neurosurgery (and surgery in general). Prior to the availability of a reliable anesthetic, time was the limiting factor in brain surgery. The surgeon was in a mad race to finish before the effects of the whiskey or other primitive and unreliable agents wore off. Since entering the brain took several hours—coupled with the additional time required to surgically alter it—neurosurgical operations at first barely advanced beyond the ancient trephinations.

Early anesthetic efforts involved such primitive measures as the use of sea sponges soaked in ether and pressed against the patient's face. Today a modern anesthetic agent acts quickly and directly on the reticular activating system in the brain stem, and the patient swiftly loses consciousness. But anesthetics, aseptic techniques, and germ-killing chemicals obviously don't perform neurosurgeries, they only make it possible for the dedicated neurosurgeon to enter into the *terra incognita* of the human mind and soul.

"Now after thou hast stitched it, thou shouldst lay fresh meat upon his wound the first day. Thou shouldst not bind it. Moor him at his mooring stakes until the period of his injury passes by. Thou shouldst treat it afterward with grease, honey and lint every day, until he recovers."

But in another instance, the papyrus deems any attempt at operation unwise:

"If thou examinest a man having a gaping wound in his head, penetrating to the bone, smashing his skull and *rending open the brain of the skull....* Thou shouldst say: 'An ailment not to be treated.'"

With the exception of trephinations, operations on the human brain itself weren't possible before the 19th century. And the first attempts at more complicated interventions into the brain were crude and fraught with peril. If patients didn't die from the operations themselves, they succumbed to postoperative infections. In addition, the methods used to relieve pain and induce sleep in the patient left much to be desired.

Alcohol, the most common choice, was not reliable; its anesthetic effects proved as variable in the operating room as in the tavern. While some

Like a good pilot, the successful neurosurgeon must know exactly where he is going and how he will get there. Also like the pilot, he must function under constraints of time, resources, and—if the operation is long and complicated enough—his own physical, mental, and emotional endurance.

The brain is too complicated and the risks of harming it are too great to justify exploratory surgery. So, prior to an operation, the neurosurgeon must precisely localize the brain site responsible for the patient's difficulties.

The first localization of record can be dated to an observation made during the Prusso-Danish War, in 1864. German physician Theodor Fritsch noticed twitching on the right side of the body of a patient who had suffered a wound to the left side of the head. Later Fritsch and a colleague, Eduard Hitzig, conducted a crude experiment on a dog, electrically stimulating the right and left hemispheres of its brain and observing that the opposite side of the body twitched. Thus they established that each side of the brain controls movement of the opposite side of the body.

The first practical application of localization took place in 1879, when Glasgow physician William Macewan encountered a young patient who had recently suffered a head injury. After some initial improvement, seizures began on the right side of the boy's face, then spread to his right arm, and finally to his right leg. Applying knowledge he had gained from the Fritsch-Hitzig experiment, Macewan opened the boy's skull on the left side. There he discovered and removed a blood clot pressing down on the left motor cortex, the brain area that controls movement of the body's right side. The boy went on to a total recovery.

Taking his head into his own hands, Peter Halvorson shows the scar of a hole he drilled to gain enlightenment. Neurologists consider such behavior a sign of illness, akin to pounding a nail into the skull in order to relieve a sense of pressure.

From this point onward, developments progressed at a quicker pace. The first neurosurgical operation for a tumor took place on November 25, 1884 at the National Hospital for the Paralyzed and Epileptic in Regent's Park, London. Neurosurgeons Sir Rickman John Godlee and A. Hughes Bennett removed a tumor from the brain of a 25-year-old Scottish farmer. Prior to the operation, the farmer had complained of increasing weakness on his left side and had suffered a series of left-side convulsions. These symptoms firmly pointed to a problem on the right side of his brain.

At the operating table, surgeons opened the skull over the right motor strip and, after piercing the dura, came upon an accessible and easily resected tumor. They removed it with a surgical spoon. Although the patient initially did well, he died within a month due to infection.

———————

Increased attention to sterile technique would cut down on deaths from infection—and would make it even more important that the neurosurgeon correctly localize his patient's problem. But even when neurosurgeons began to learn the exact sites of what they called lesions, their success rate was nothing to brag about; more than half of their patients died. By today's standards, brain surgery was an ugly, dirty, and very risky enterprise during its first 25 years. All that changed in 1910, with the brain-tumor operation carried out by a young neurosurgeon named Harvey Cushing, on General of the Army Leonard Wood.

As with many neurosurgical operations today, Cushing carried out the procedure with his patient fully awake. This is possible because, while the coverings of the brain contain pain receptors, the brain itself does not. Thus, a local anesthetic is all that's needed to ensure a pain-free operation.

Despite the conviction of relatives that General Wood was as good as dead (one of them after briefly visiting the operating room referred to the general in the past tense), Cushing's famous patient went on to make a complete recovery. This widely trumpeted success helped establish Cushing as the most accomplished neurosurgeon in the world. Over his career, his operative mortality rate was less than 10 percent, an incredible statistic for his day.

In addition to his surgical skills, Cushing introduced technical innovations such as the use of "ether charts"—operating recordings of the patient's blood pressure and pulse rates during surgery. He developed mechanical aids that included a special tourniquet to control bleeding, various clips for application to blood vessels, and the electrocautery, a device that controlled bleeding by singeing the tips of small arteries. Even by today's standards, Cushing was a remarkable man. Along with publishing in five volumes the 2,000 confirmed cases of brain tumors he had personally operated on, Cushing was able to hone his literary talents. In his spare time away from the operating theater he wrote a popular biography that won him a Pulitzer Prize.

Starting with Cushing, technology became increasingly important to neurosurgery. A number of advances stand out:
• The development of the CT scan (computer-assisted tomography) enabled doctors to make early diagnoses of brain diseases. CT scan images offered a powerful advantage over conventional x-rays, for they provided "slices" of the brain at various depths; the brain could be peeled like an onion and studied

at different levels. They also provided images in different planes and from almost any conceivable point of view. Today, the "cuts" can be made finer and finer, in a process similar to what happens when you cut thinner pieces of salami by using a sharper knife with a narrower blade.

• MRI, or magnetic resonance imaging, employs radio-frequency pulses of energy and a strong magnetic field, rather than x-rays. MRI renders brain anatomy in greater detail, and thus detects many abnormalities that elude CT scans.

• While both CT and MRI provide excellent images of brain structure, they determine nothing about brain function. That's where functional magnetic resonance imaging (fMRI) and magnetic resonance angiography (MRA) excel. MRA offers high resolution, depicting brain structures that ordinarily are visible only by microscope.

• But the most revolutionary innovation leading to the modern wonders of neurosurgery occurred in the mid-1960s, when neurosurgeon Gazi Yasargil introduced the intraoperative microscope. This innovation enabled neurosurgeons for the first time to visualize tiny blood vessels and nerve endings branching from their major sources. Learning to use the microscope required major revision in terms of the neurosurgeon's techniques and his relationship to the brain.

Instead of looking directly at the brain, the neurosurgeon stares into a microscope that greatly magnifies everything in the operative field. A nerve filament thinner than a human hair takes on the dimensions of a garden hose, while scissors small enough to fit into the hand of a doll look like hedge clippers. Throughout each operation the neurosurgeon has to correlate the dimensions of the surgical instruments as they appear through the microscope with how they feel in his hands.

• Three-dimensional television has taken the operating microscope even further. Until its invention, neurosurgeons-in-training observed operations by taking turns at the neurosurgeon's side to peer through a second binocular lens attached to the intraoperative microscope. While this was tremendously instructive—the only way to appreciate the richness of the three-dimensional operative field—it could accommodate only one observer at a time. Others were restricted to a conventional, two-dimensional television monitor. That image lacked depth, depicting everything like a photograph from an anatomy book. Useful, of course, but lacking the critical third dimension in a world where life and death can be measured in fractions of a millimeter.

The combination of computers and high-resolution imaging devices has so radically changed neurosurgery that, were Harvey Cushing alive today, he would surely pause in awe at least momentarily. But you can bet that, an instant later, he would be using these new innovations to carry out operations that would have seemed like science fiction in his time.

For instance, the treatment of brain tumors—Cushing's specialty—now involves highly focused beams of radiation. Instead of using steel scalpels to cut into and inadvertently disrupt or harm delicate and vital structures such as the optic nerves or the carotid arteries, today's neurosurgeon employs finely focused radiation to sear tumors into oblivion.

First, of course, he must determine the exact site to focus the killing beams. This is all part of what's called gamma knife radiosurgery. A head

High theater of surgery: Nearly a century ago, the operating room at Johns Hopkins Hospital provided plenty of seating for medical students to observe basic procedures before joining master surgeons like Harvey Cushing at the operating table. Today, three-dimensional television monitors, operating microscopes, and other high-tech devices enable tomorrow's neurosurgeons to see the current masters of the craft perform, in intimate detail.

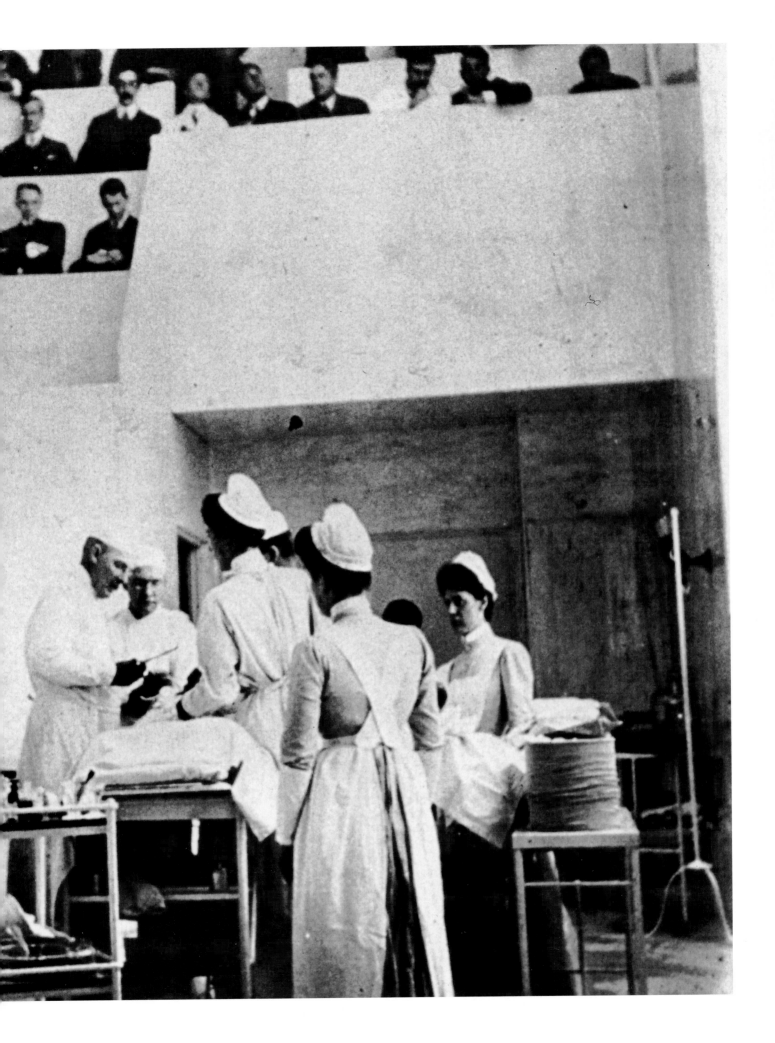

frame is affixed to the patient's head. Next, the neurosurgeon obtains a set of localizing MRI or CT images to identify targets within the brain. A high-speed computer then analyzes the scans and provides precise coordinates for the diseased area to be irradiated. Finally, a small conveyer belt brings the patient into position within a dome that contains 201 individual radioactive cobalt sources, positioned to focus gamma rays at a given spot with a treatment accuracy of within 1 millimeter. The process involves no surgery in the common definition—that is, surgery with steel tools. The patient experiences no pain, can be allowed to walk about later that same day, and is discharged the next morning.

There are limitations, however. Because the gamma knife delivers high radiation doses in a single fraction of a second, it doesn't work well with large tumors that infiltrate normal brain tissue like ink spilled onto an oriental carpet. Such diffuse targets are simply too imprecisely localized for the gamma knife's finely honed focus. In such instances, traditional brain operations must be employed either prior to or as an alternative to radiation treatment. Computers can help here too.

———————————

The viewing wand, a creation of ISG Technologies in Toronto, enables the neurosurgeon to reconstruct on a computer screen a three-dimensional model of the brain and the operative target. The instrument consists of a computer and a mechanical wand. Initially, the computer displays a three-dimensional image of the patient's head, based on CT and MRI scans previously loaded into its memory banks. As the neurosurgeon touches different points on the patient's scalp with the wand, sensors at the tip convey information to a computer that synthesizes it. Next, the neurosurgeon locates the same reference point displayed on the computer monitor. Each time he matches a landmark on the patient with the same landmark on the computer screen, the computer adjusts its image. When a sufficient number of points are correlated, the neurosurgeon can simply move the wand over the patient's head and view on the screen a detailed image of what lies beneath.

At this point the neurosurgeon can engage in a mock operation. By manipulating the computerized image into different screens that measure volume, distance, and angle, he can "cut" into the imaged tumor and see the results of his actions simulated on the computer. An alternative attack from a different direction can then be mounted and compared with the first approach.

"Computer technology takes a lot of the guessing out of what you're doing," attests Benjamin Carson, director of pediatric neurosurgery at Johns Hopkins Hospital. "When you can practice going precisely to a place without poking here and there, the exposure of the patient decreases significantly."

Thanks to computers and other technological advances, few of the earliest tools of neurosurgery are still in use today. After the initial incision, the neurosurgeon abandons the traditional scalpel for high-tech tools such as the Midas Rex drill, electric cautery, and imaging devices like the intraoperative microscope with its built-in wand and controls for video and still photography. As the neurosurgeon deftly maneuvers into the interior of the patient's brain, his every movement can be projected from the operating room into an auditorium filled with hundreds of neuroscientists-in-training who watch the action with the aid of special 3-D glasses.

What they see is wonderful and endlessly fascinating. Journalist Edward J. Sylvester, in his monumental book, *The Healing Blade*, captures the beauty as well as the science involved when a high-resolution, three-dimensional microscope delivers to the monitor the sense of walking into the brain itself. Comparing the experience to an airplane's flight, Sylvester writes:

"The relativity of depth and distance requires contrast for the mind to grasp what the eye sees. Fly at 30,000 feet on a clear day and you see only a flat, colored landscape at an indeterminate distance below. Then if you enter billowing trees of cumulus clouds and, through a sudden parting that reveals a lake of air, see the earth 30,000 feet below, it is a stunning vision.

"Each time you follow this journey down, you see things you didn't see before, within structures or wrapped around them or nested within, but it's never quite the same as what you saw before. It's all new and rises right before your eyes; that is the power of resolution. There is no end to it."

The neurosurgery of the new millennium promises to be equally exciting. Treatments for Parkinson's and maybe even Alzheimer's diseases will involve direct delivery of deficient or missing chemicals to precisely selected targets. Advances in genetic engineering will further the development of what is already being called molecular neurosurgery.

For example, work is already underway to develop genetically engineered viruses that will be able to kill deadly tumor cells while exerting no influence on normal brain cells. When such vehicles become available, the neurosurgeon will insert the virus in the proper location within the brain and things will work pretty much automatically from there. Drugs for epilepsy and other neurological conditions will also be delivered to the proper targets, thereby cutting down on seizure frequency and perhaps eliminating the seizure "focus" altogether—to say nothing of reducing unwanted side effects. As an aid to all these procedures, virtual-reality programs will allow the neurosurgeon to practice—long before he or she ever enters the operating room.

"Links to large databases will provide physicians with instant access to videos of surgical procedures," according to James I. Ausman of the Department of Neurosurgery, University of Illinois at Chicago. Ausman also predicts that the operating microscope will be "replaced with fiberoptic television cameras, which are directed by the surgeon's hand while the surgeon watches the progress of the procedure on a monitor. Ultimately, robotic surgery will follow."

It's expected that these new robotic tools will wend their way through the brain with a precision that will amaze even the most skilled human operator. But however technology changes, the basic mission of the neurosurgeon will remain the same: to apply his marvelous skills against the most dangerous and destructive illnesses that plague us, illnesses of the organ responsible for all that we are.

To me, an Emily Dickinson poem continues to capture the essence of the privilege and challenge that face every neurosurgeon:

Surgeons must be very careful
When they take their knife!
Underneath their fine incisions
Stirs the Culprit—Life!

■ SECTION V

FUTURE
DIRECTIONS

FUTURE DIRECTIONS

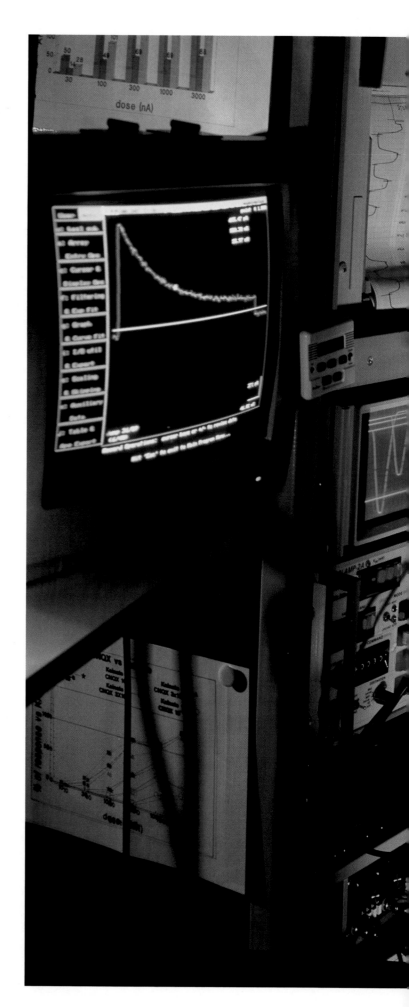

Crystal balls never come with a guarantee—quite simply, because no one can predict the future. Certainly nobody predicted the advances in our understanding of the brain that have occurred during the past two decades. While that might serve as a warning to the wise, we now know enough about the brain to enable us to discern, at least dimly, some aspects of neuroscience's future. With these caveats in mind, let's explore some possible new directions and sample some of the thinking from neuroscientists on the cutting edge.

Floyd Bloom, with the Scripps Research Institute's Department of Neuropharmacology, believes, "It's likely there will continue to be many new neurotransmitters discovered and many new forms of drug-brain interaction developed as we learn more about the systems affected in, for example, depression, schizophrenia, or substance abuse."

Such advances will involve identifying specific molecules that bolster the actions of neurotransmitters. Such identifications will aid in the development of new drugs with selective actions.

■ Developing new drugs for brain-related illnesses means powerful computers, sophisticated instruments—and dedicated researchers like this scientist at the pharmaceutical giant Glaxo-Wellcome. PREVIOUS PAGES: Just as hearing aids give these Israeli children the ability to detect sound, computerized and brain-activated prosthetic devices of the future hold the promise of a better life for amputees and others with disabling injuries.

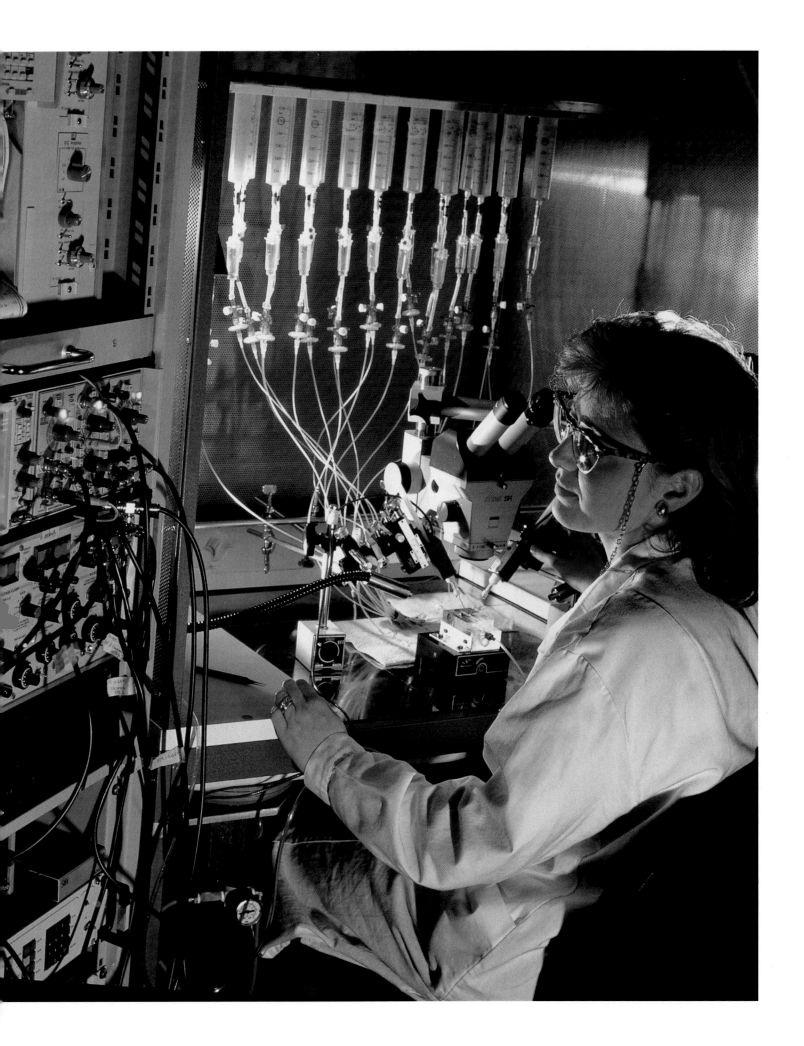

One example is the neurotransmitter GABA. After neuroscientists noted a deficiency of GABA in certain forms of epilepsy, they were able to devise anti-epileptic drugs that either targeted the enzymes that make GABA, the proteins that bind to it, or the enzymes that break it down into its component parts. Thanks to these advances, treatment of epilepsy moved from blind empiricism—try a little of this or that and see if any of it works—to the design of specific medications capable of influencing a selected neurotransmitter.

A similar process likely will underlie the development of new drugs to treat schizophrenia and depression. At the moment, the prime emphasis concerns dopamine, serotonin, and epinephrine. Yet according to current estimates, more than a hundred different chemical substances within the brain may function as neurotransmitters or neuromodulators. It's likely that some of them—acting alone or, more likely, in concert—play important roles in normal and diseased states.

"The belief that the complex cognitive and emotional states that underlie any emotional disorder are regulated by a single neurotransmitter is probably no more valid than the idea held earlier by phrenologists who believed that complex mental attributes could be localized in one specific part of the brain," writes Elliot S. Valenstein, emeritus professor of psychology and neuroscience at the University of Michigan.

We can look forward not only to the discovery of new neurotransmitters but also to new ideas about how those neurotransmitters operate. A drug may influence one or more neurotransmitters within hours—yet its beneficial effects may be delayed for weeks. How is this possible?

As mentioned in the section on addiction, the binding of a neurotransmitter to its postsynaptic receptor can set off a chain of biochemical changes that culminates in the production of an intracellular second messenger. This novel chemical takes aim at a wide variety of different targets.

"There is a cascade from cell surface receptor to the genetic program within the cell," according to Robert J. Birnbaum, a psychiatrist associated with Harvard Medical School. This cascade is influenced not only by neurotransmitters but also by hormonal molecules that can penetrate the neuron's outer membrane and then influence DNA regulatory elements within the neuron's nucleus.

Each influence on the neuronal nucleus has the potential to activate genes responsible for specific behaviors. Some of these influences take place over the space of a few hours, perhaps even minutes—while other responses may be delayed for decades. Thus a drug may exert an immediate effect on neurotransmitters but a time-delayed effect on behavior. This delay represents the time required for chemicals to influence the neuron's nucleus. An arrangement of multiple factors affecting DNA at different times and locations in the cell nucleus provides a rich variety of potential responses.

———————————

Whatever discoveries will be made in the future concerning the action of neurotransmitters within various regions of the neuron, mental illnesses will be increasingly understood and defined in terms of brain functioning. For example, as noted earlier, serious depression and post-traumatic stress disorder (PTSD) shrink the hippocampus (that small, seahorse-shaped structure important in the first stages of memory formation). Newly emerging evidence suggests that even milder depressions, from which the patient recovers completely, may leave a permanent imprint on the hippocampus.

Using MRI, scientists at Washington University, in St. Louis, discovered that healthy women with a history of depression had significantly smaller hippocampal volumes compared to women who had never experienced depression. The most likely explanation for this difference involves brain-damaging stress hormones such as glucocorticoids, which are known to shrink hippocampi in stressed individuals. Thus rapid and effective treatments for depression have the potential not only for relieving the depression-associated suffering, but also for protecting the hippocampus from further damage.

In addition to depression, it's expected that other conditions, such as panic and generalized anxiety disorder (GAD), obsessive-compulsive disorder (OCD), and schizophrenia will be more firmly established as brain disorders in the near future. Treatments will aim at both relieving mental distress and restoring healthy brain functioning.

Also looming on the horizon are critical growth factors that will provide mechanisms for neuronal regrowth or self-repair. Neuroscientists already can slow down the aging of the brain in monkeys by implanting cells that secrete nerve growth factor, or NGF. The NGF increases cell division in the selected area and thereby reverses age-associated brain changes. One area, a thin strip of cells known as Ch4, is particularly important.

As monkeys age, Ch4 cells suffer a decrease in a key chemical called p75, abundant in the brains of younger monkeys. But if cells secreting NGF are implanted near the Ch4 cells, the levels of p75

increase. Three months after implantation, the size of the Ch4 cells and their p75 production just about equal the levels found in younger monkeys.

Will this change in brain typography lead to a reversal of mental decline in monkeys? So far, we don't know. But even if it does, there's no guarantee that similar results would occur in humans. And even if *that* were the case, a more practical delivery system for NGF would have to be found, since cell implantation is both risky and impractical. Rather than bathe the whole brain in NGF, we would want to pinpoint its delivery to the precise sites where it's needed. One possible route to that end would be to modify specific genes.

"As the major brain genes are identified and neurotransmitters and their circuits are known, it is quite likely that some cases of missing or excessive neurotransmitters can be solved by drugs that will turn on or off the appropriate genes in the crucial locations," says Floyd Bloom.

Over the next decade, neuroscientists expect to make new discoveries about behavior, thanks to genetic research on what are known as knockout mice. The term refers to genetic modification; a knockout mouse is one that has been genetically engineered to prevent the expression of a specific gene. The first generation of knockouts prevented the expression of a factor thought to be necessary for long-term potentiation (LTP), a special form of memory. Later refinements enabled neuroscientists to produce a mouse that lacked one form of the neuronal receptor NMDA, but only in a specific area of the hippocampus known to be important in the formation of memory for spatial location. As a result, firing rates for neurons in this area decreased; the mouse got lost more easily and seemed unable to form functional spatial maps. Future research on knockout mice will help neuroscientists bring together diverse findings and ideas drawn from molecular, cellular, and behavioral levels.

One thing is certain: Awaiting discovery are an extremely large number of genes that contribute to normal and abnormal brain functioning. Some of these genes may influence the formation of proteins that regulate subtle brain processes. Some may control the inherent excitability of certain neurons; other genes may control the tendency of neurons to form networks. The discovery of such genetic programs could shed light on why experiences influence the behavior of some people but not others. Research might even provide a genetic "fingerprint" enabling scientists to make a rough estimate of the likelihood that a particular person might find an experience stressful or invigorating or unpleasant.

As vital as genetic inheritance is, it is only part of the human story. When an identical twin has manic depression, the chance of the other twin also having the disorder is not 100 percent but 60 to 80 percent. (If the twins are fraternal—that is, they arose from two eggs rather than a single egg—the odds of both being affected drop to only 8 percent.) The fact that identical twins have a risk of 60 to 80 percent—rather than 100 percent, due to their identical genetic composition—represents the contribution of the environment.

Often that contribution comes from what scientists call a "second hit." If you carry a gene that predisposes you to lung cancer, you may never

develop the disease—if you don't smoke. But if you're a smoker, that habit might serve as an environmental second hit capable of converting your genetic vulnerability into outright disease. Another person lacking that genetic predisposition may smoke for decades and not get lung cancer.

Until we know more about our own genetic predispositions, the safest approach is to assume that we are predisposed towards illnesses stemming from harmful and potentially dangerous activities. We all know that life isn't always fair—and much of that unfairness stems from the different cards we all carry within our genetic deck. With future genetic discoveries, it's likely that each of us will learn more about the genetic cards we carry; that knowledge will help us make increasingly informed decisions about how to live our lives.

Undoubtedly, future advances in brain research will be linked to new developments in technology. Prior to the invention of CT and MRI scans, we could only guess about the functioning of the living brain. New and different technologies will deepen our understanding of the brain even further.

Already, there have been surprises. Vision, for instance, turns out to involve widely separated but closely interlinked brain areas. Electrodes placed into individual brain cells of monkeys reveal that their visual system contains at least 34 maps of their visual world; each analyzes a different aspect, such as color or orientation. A similar vision-processing system is believed to hold true for humans.

These visual maps are arranged in a hierarchical manner, with signals radiating along two main pathways from the primary visual receiving area at the back of the brain, in the occipital lobes. A "what" pathway extends from the primary visual cortex into the temporal lobes and specializes in things like shape, color, and speed of movement. A "where" pathway, ascending from the primary visual pathway into the parietal lobes, is sensitive to spatial location. Of course, each of these specialized areas can be drawn upon to provide information about the other; the shape of an object may offer a clue to its position. As in a symphony, every instrument has the potential to influence the others.

Imaging studies have also provided new insights into language. Until the advent of PET scans, neurologists believed that language processing was confined to the classical areas first described in the mid-19th century by Broca and Wernicke. New research has improved on that oversimplified concept. A number of additional areas, mostly in the left hemisphere, are now known to cooperate in fashioning different aspects of language, such as word choice and grammar.

"The vast lexicon of words we use in everyday language is not based in one single brain center or region, but depends, rather, on separate neural systems which are relatively specialized in the processing of different words," writes Antonio R. Damasio. A neurologist at the University of Iowa College of Medicine, Damasio and his wife and fellow neurologist, Hanna, are responsible for pioneering work with CT and MRI that elucidates the brain mechanisms underlying psychological processes.

The Damasios and other researchers have turned up some real surprises. For instance, verbs are located in the parietal and frontal regions, while nouns are distributed in the occipito-temporal regions. Even finer distinctions occur: Nouns for tools or animals or specific individuals are based on different neuronal networks, so if one network becomes damaged, a category-specific disorder can result, causing a person difficulty in understanding colors, letters, numbers, or other categories.

For instance, a patient described by the English neuropsychologist Elizabeth Warrington experienced no difficulty identifying inanimate objects but was greatly impaired when confronted with a living thing. A thermometer was correctly defined as "a device for registering temperature" while a butterfly was referred to as "a bird," a frog as "an animal, not tamed," and a camel "not recognized at all."

At the moment, neuroscientists know for certain that different aspects of language are processed in different brain areas—and that these differences can be imaged. Different areas become active when a test subject successively views a word, listens to it, speaks it, or generates a verb in response to it. The brain also uses different pathways when attempting responses to familiar versus unfamiliar lists.

Functional imaging techniques such as PET and fMRI promise to provide personal brain maps that will be more individuating than fingerprints. But unlike fingerprints, such brain maps can be altered. Their malleability has important implications. We know, for instance, that we can change important aspects of our brain's organization and structure. And a brain map taken at any given moment has the potential to enable each of us to discover our brains' assets and liabilities.

Some people are good at mathematics; others can't seem to carry out the simplest computation. Soon we will be able to understand these differences in terms of brain organization and functioning. This could lead to redesigns in educational methods and

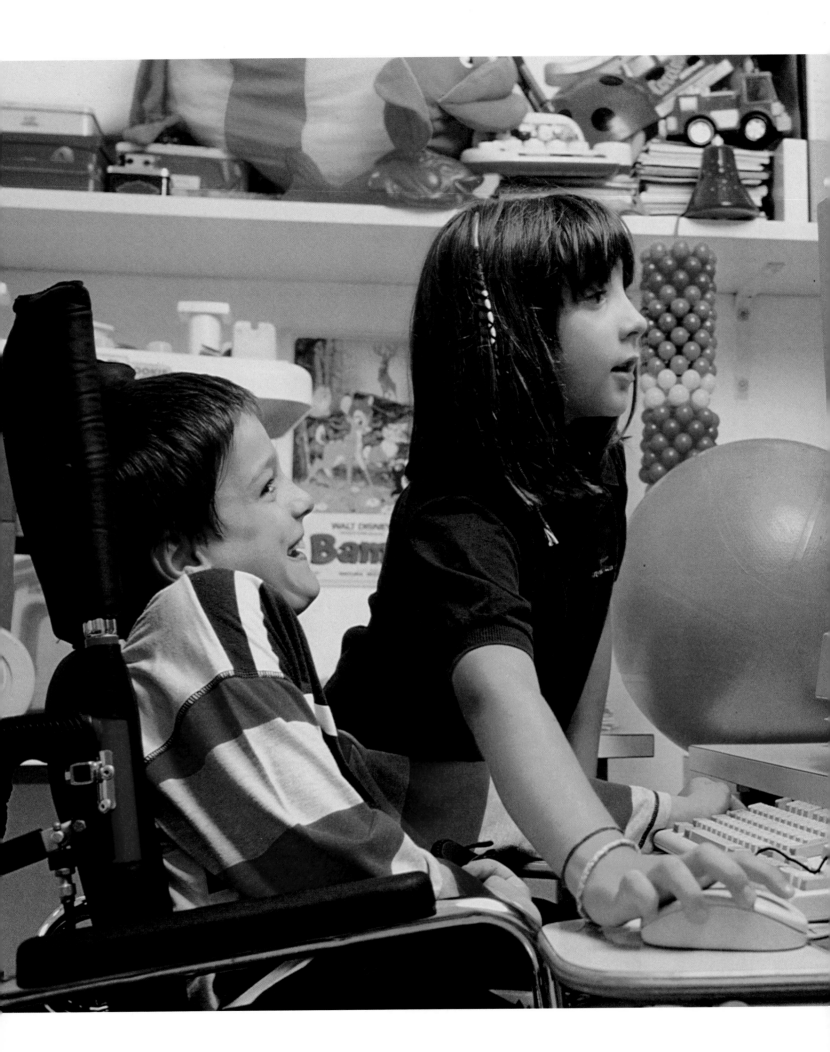

RESTART
OPTIONS
MINDDRIVE
HELP
SAVE
QUIT GAME
30:00
PHILIPS

■ High-tech magic empowers this disabled boy to play a computer game with his friend. A sensor on his finger measures mental activity by monitoring electric signals in the skin as well as heartbeat and body temperature. Software filters out unconscious nerve impulses and focuses on conscious ones, thus making this computer truly brain-activated. Similar technological improvements can help some deaf and blind people recover at least part of their hearing or sight. Even so, the human brain remains far more complicated than any computer.

even new ways of defining intelligence. Already psychologists have largely changed their minds about the value of standard IQ testing. They have encountered too many people who, despite average performances on standard intelligence tests, have gone on to creative achievements. The opposite also holds: Many children who showed extremely high IQ when tested in the 1940s by psychologist David Terman turned out to accomplish little as adults.

Future imaging applications will also help explain some of the structural and functional differences between one brain and another. For instance, it's been known for years that people who are born deaf and learn to communicate by sign language develop greater peripheral vision than those who don't. PET studies provide an explanation: the visual orienting system located in the parietal lobe of the deaf person expands at the expense of areas that in a normally hearing person would be devoted to sound.

Such findings raise several intriguing questions. How does the brain of the architect—with her enhanced ability to image structures strictly in the imagination—differ from that of the lawyer who fashions arguments in the interests of a particular point of view? How does the brain of the artist differ from that of the novelist? Imaging will help answer these kinds of questions. Indeed, cognitive neuroscience—the study of every aspect of human thinking—will rapidly expand in the early years of the 21st century.

Vernon Mountcastle, an emeritus professor of neuroscience at Johns Hopkins University, writes, "Imaging studies have already shown that primary sensory areas vary in size between individuals and between hemispheres in the same brain, and that they can be changed in size by intensive sensory or motor experiences."

Methods of imaging the brain will also make significant contributions to our self-understanding. They should be particularly helpful in revealing the processes that accompany consciousness. Soon we should be able to image the conscious brain and correlate those images with timed measurements of the brain's electrical activity.

"We are being provided with much new data about how our organ of cognition is functionally organized," adds professor Richard S. J. Frackowiak, at the Institute of Neurology, in London, England. "Our brains are now able to reflect upon themselves using rigorous scientific methods and instrumental forms of measurement. These new imaging methods are leading to better human self-understanding through the appreciation of the unique nature of how our mental activity is implemented in the organic matter that we carry around in our skulls, that enormously complex organ that defines our personalities, hopes, wishes, actions, and ambitions."

Zach W. Hall, vice chancellor for research at the University of California, agrees. "Brain science in the future will be shaped by new technologies, technologies that are difficult to predict, and in some cases, even to imagine. We have already reached the stage where the complexity of neuronal networks exceeds our intuitive understanding. Even with present-day computers we simply can't encompass that degree of complexity. In the future we will need new methods of analysis."

One promising method involves simultaneous recording from groups of neurons, using multiple electrodes in special arrays. Coupled with this will

Brain science in the future will be shaped by new technologies that are difficult to predict, and in some cases, even to imagine.

—Zach W. Hall

be improved detection and correlation of subtle electrical signals. This will require increasingly sensitive recording devices, and computers capable of handling massive amounts of data. All of these will help overcome an inherent and fundamental limitation of the brain—the number of elements it can consider at any given time—in order to understand the brain's own functioning.

On a practical level, imaging techniques may help distinguish one mental illness from another, serve as a measure of treatment success, even predict who will respond to a particular treatment. Already, functional MRI provides a window on how well a patient with obsessive-compulsive disorder responds to a particular drug; successful therapy brings about significant changes in the fMRI activity of the caudate nucleus, while those who fail to respond to the treatment show no such changes.

Panic disorder can be diagnosed on the basis of PET studies. A specific PET signature marks each panic episode: decreased blood flow to parts of the frontal and temporal lobes, the anterior cingulate, and certain limbic areas. This holds true for people who have been diagnosed with panic disorder, as well as those not usually affected by panic.

Children with attention-deficit hyperactivity disorder (ADHD) exhibit different activation of brain regions than normal children do. But after treatment, the brain-activation patterns of ADHD children shift towards normal.

Robert C. Malenka, a psychiatrist at the Stanford University School of Medicine, believes, "Although advances in human brain imaging are just beginning to be applied to psychiatric illnesses, they have the potential over the next decade to revolutionize our diagnosis and treatment of these devastating disorders. If Freud were alive today, he would probably be using these techniques to probe the unconscious."

Indeed, if Freud were attempting deeper levels of analysis, he would find extraordinary progress in structural biology, the visualizing and understanding of brain structures in three dimensions. Thanks to this progress, neuroscientists now hope to identify new ion channels, new neurotransmitters, and new receptors, all over the next few years. From such discoveries will emerge novel drugs, along with an enriched understanding of how nerve cells grow, migrate, and enter into circuits and other linkages.

New technologies will improve both the diagnosis and the treatment of neuropsychiatric illnesses. Brain injury will no longer be considered hopeless. First hints that the brain can repair itself date back about a century, when active animals were observed to have a better chance of recovering from brain injury than inactive ones. In humans, the forced use of an injured arm or leg was a popular treatment for a while, but it fell out of favor early in this century, when the crude measuring equipment of the time failed to document any changes.

A renewed interest in brain rehabilitation emerged in the 1980s and 1990s, based on the plasticity model of brain function and repair. Researchers discovered that the sensory and motor cortical maps for the hands display huge areas of overlap. In addition, they observed that brain maps program the movement of an entire limb, not just movement of individual muscles or joints in the arm. Even more exciting was the finding that training can reshape those maps.

New generation of aptitude tests comes to kindergarten in this San Jose, California, school. Spawned by new brain-related technology, such tests seek to measure cerebral responses directly, thus eliminating cultural bias, which colors the results of some conventional tests.

While successful training can require an enormous number of repetitions, programming those repetitions comes easily to the brain. The injured person must persevere with rehabilitative efforts; the process isn't any different from what happens when someone wishes to learn to play the piano at concert level or hit a baseball with the skill of a major-league player. Such goals require the activation of millions of muscle cells, carried out over many years. Rehabilitation uses repetition to enhance the functioning of normal brain cells and thereby compensate for lost and damaged ones.

"Cortical injury has an inevitable effect on the function of adjacent intact tissue. Skilled movement also has an effect on cerebral cortex. After injury these two variables interact," observes Randolph Nudo, of the University of Kansas Medical Center. Drugs can also enhance rehabilitative efforts. For instance, amphetamine improves brain plasticity, and may become a standard treatment for brain damage. Several researchers have argued that amphetamine appears to be effective only when combined with physical training.

New advances in the treatment of neuropsychiatric diseases are also just around the corner. At the moment, depression is diagnosed on the basis of a patient's symptoms and complaints. Depending on what the patient says, the psychiatrist makes very general distinctions, such as bipolar (manic depression) or unipolar (depression alone). Drug treatments are then decided upon without specific reference to brain functioning. Typically, if one drug proves ineffective, the psychiatrist turns to another. While not exactly haphazard, this approach is fairly crude compared to treatments for, say, infectious diseases, in which the physician concentrates on the identification of the causative infectious agent, its mechanisms of attack, the body's defenses at various levels, and the biochemical changes characteristic of a particular illness.

Current approaches to the diagnosis and treatments for epilepsy provide a model for possible future directions in our understanding of depression. Not long ago epilepsy, like depression today, was considered a single illness resulting from diffuse brain disease. But thanks to electroencephalography and neuroimaging, neuroscientists now speak not of "epilepsy" but of "the epilepsies"—many different forms of electrical malfunctioning that originate in many different areas of the brain. Technological advances now empower neuroscientists with detailed information about what brain areas are dysfunctional in an epileptic, the cause of that dysfunction, and the treatments most likely to prove effective.

We also can look forward to a transformation in our understanding of what we might call "the depressions." Studies using PET and SPECT (single photon emission computed tomography—similar to PET but somewhat less powerful) are already in use as a means of distinguishing the depressive component of manic depression from unipolar depression. Scanning also helps in identifying those depressed patients likely to respond to Prozac or other drugs. Imaging studies can even provide an early warning of relapse in successfully treated patients.

Included among the most promising treatment techniques are:
• Transcranial electrical stimulation. This involves passing of low-level electrical energy over the skull. The skull acts as a large resistor and redirects the applied current through many brain areas. So far,

the method has proven effective in the treatment of Parkinson's disease and alcoholism.

• Transcranial magnetic stimulation (TMS). A large magnet is held against the scalp to induce electrical currents in the underlying cerebral cortex. This depolarizes the cortical neurons, which then influence the firing patterns of neurons farther downstream within brain regions connected to the cortex. Applied to the proper prefrontal area, TMS has elicited positive mood changes in normal volunteers. The hope is that it will do the same for people diagnosed with depression, by inducing long-term changes in key brain circuits. So far, trials have produced mixed results.

Strangely, we know more about diseased brains than we do about healthy ones. This is due to laudable efforts over the years to cure brain diseases or at least enhance the lives of people stricken with illnesses for which a cure didn't exist. In addition, many early forays into the *terra incognita* of the brain could not be used in healthy people, because the instruments and methods involved excessive risks. And while curing or ameliorating a disease may justify some controlled risks, incursions into the normally functioning brain does not. But with the introduction of essentially risk-free technologies like MRIs, neuroscientists can now enrich our understanding of how the normal brain works.

As an example of how things are likely to evolve in the future, consider some recent findings about the brain's ability to anticipate. In 1992 Joaquin Fuster and his associates of the UCLA School of Medicine discovered a special network of neurons in area 46 of the monkey prefrontal cortex. These cells increase in activity during those few seconds between a stimulus—such as a flash of colored light—and the monkey's trained response to move either the right or left forelimb. In contrast, activity decreased in a second group of cells in the prefrontal cortex, which respond directly to the color of the light stimulus. Fuster believes the first group may be responsible for planning and the second group for memory. He calls them "cells that look to the future and cells that look to the past."

Such future-looking cells also play a role in our anticipation of pain. Our brain's ability to anticipate pain serves useful purposes. We can predict that cutting our finger while dicing onions will prove painful, so we proceed cautiously. But there's a downside to our ability to anticipate pain. Often our anticipatory fears are worse than the actual pain involved. (Think of the dreadful anticipation that often precedes a visit to the dentist.)

Recently, neuroscientists discovered the brain areas that are activated when we become anxious or fearful about an impending painful experience. This expectation of pain activates key sites within the medial frontal lobe, the insular cortex, and the cerebellum, at locations that are close to but distinct from areas activated when we actually undergo a painful experience. This finding helps explain why tranquilizers and sedatives can prove effective: They modify activity in this so-called pain anticipation circuit. Newer drugs are currently under development that will selectively hook onto cell receptors in the medial frontal/insular/cerebellar circuit.

In a related development, studies of musicians and athletes show that simply imagining a complex or skillful movement can aid its performance. The brain areas involved in imagining specific motor

Bounty of the rain forest—home of many pharmacologically active plants—may someday include drugs for brain-related illnesses. Currently, major pharmaceutical firms have begun cooperative ventures with indigenous peoples to decipher the chemistry of plants favored by native healers. Growing enthusiasm, worldwide, for herbal remedies such as ginseng and St. John's wort attests to the increasing willingness of both patients and doctors to consider new approaches.

movements surround those areas that are associated with actually making the movements. So when you imagine yourself serving a tennis ball, you activate a number of brain regions in front of and behind the primary motor areas. These "future-looking" cells are thought to program the movement; they fire prior to the firing of cells in the motor cortex that actually activate the appropriate muscles in your arm and hand.

"Such simple experiments illustrate dramatically that it is possible to image brain activity associated with 'pure thought,'" believes professor Richard S. J. Frackowiak. "Hence the idea that thought and introspection can be physiologically studied is, at least in principle, realized."

In the near future, fMRI, PET, and other imaging technologies promise further insights into the neurological basis for mental operations. We already know that many of these areas lie within the frontal lobes, areas that are especially large in the human brain and are involved in circuits that underlie thoughts, aspirations, future plans, and other distinctly human mental processes.

"A description of the physical basis of pure mental activity, even with the present degree of precision, is astonishing and poses challenging questions to philosophers interested in the nature of the 'mind,'" adds Frackowiak.

For starters, functional imaging may provide an acceptable alternative to IQ tests as indicators of intelligence. As critics correctly point out, standard IQ tests often underestimate the intelligence of culturally underprivileged children and adults. But imaging devices like PET or fMRI show differences in brain activation that correlate with intelligence as measured by other tests. Subjects taking the Ravens test (a more culturally unbiased measure of IQ and other mental traits) display different PET scan patterns depending on their IQ.

In addition to intelligence, other aspects of learning will be affected by emerging imaging technologies. Already dyslexia—often distinguished by problems with learning to read, erratic spelling, and other deficits primarily affecting written as opposed to spoken language—can be diagnosed with the help of images showing the brain structures important in normal reading. In dyslexics, the angular gyrus—an important area toward the back of the brain—seems to function abnormally during reading. To compensate for this failure, dyslexics often recruit other brain areas, such as the inferior frontal gyrus, located toward the front of the brain. In normal readers the inferior frontal gyrus is associated with spoken rather than written language; with dyslexics the pattern is different. Specialists speculate that dyslexics who say the words they are reading under their breath may be relying on the inferior frontal gyrus rather than the angular gyrus.

With imaging studies, treatments for dyslexia can now be evaluated on the basis of whether or not they nudge brain activity in the direction of a normal reading pattern. The goal is a smoother integration of the brain systems that process written language, the speech-sound processing system, and the visual system. Such efforts are an early venture into revealing the neurological mechanism underlying many aspects of normal human mental activity.

As part of this venture, neuroscientists within this decade will come up with neural prosthetic devices that will enable blind persons to perceive

Dyslexia renders reading a lifelong challenge to otherwise able individuals. In dyslexics like this girl, an area toward the back of the brain vital for reading skills seems to function abnormally. Current treatments seek to foster greater integration of those brain systems responsible for written language, the speech-sound processing system, and the visual system.

shapes such as squares and letters, as well as devices that can help the deaf hear and understand human speech. In fact, helping the deaf has so far proved much easier than restoring sight to the blind. Over a million nerves in the eye carry signals to the brain from light receptors in the retina. At the moment, there is no way of stimulating all of those nerves independently. The number of receptors in the ear is a thousand-fold less, so a lot can be accomplished using a small number of electrodes.

The cochlear prostheses already on the market enable the user to pick up enough sound to engage in conversation. These devices consist of a small microphone outside the ear, which picks up sound and passes it along for analysis to a signal processor. This apparatus breaks down the complex sounds of human speech into their constituent tones—an operation normally carried out by the cochlea. The electronic transformations of the various tones are then transmitted to an antenna implanted beneath the skin behind the ear. Minute wires wend from the antenna into the cochlea. There, a surgically implanted array of electrodes stimulates clusters of auditory nerve fibers, activating them in a pattern very similar to what happens in the cochleas of people with normal hearing.

"The impact of the prosthesis is enormous," according to A. James Hudspeth of the Howard Hughes Medical Institute and the Laboratory of Sensory Neuroscience at Rockefeller University. "For the first time, wholly deaf individuals may regain the ability to enjoy speech and music."

Each cochlea normally contains about 16,000 sensory receptors, known as hair cells. As we grow older, we steadily lose our hair cells. This loss is the eventual cause of deafness among those who

gradually become deaf. But there is some good news: Hair cells can regenerate in frogs, chickens, and other species, so why not in humans?

"Several factors favor this enterprise," explains Hudspeth. "First, our own cochleas contain supporting cells like those that can regenerate hair cells in other species. If these supporting cells could be coaxed to divide and mature in humans, they could repopulate a cochlea that has lost its receptors. Just as the cochlear prosthesis represents the most successful technological solution to date for a neurological problem, hair-cell regeneration has an excellent chance of providing the first cellular rejuvenation of an essential part of the human nervous system."

But the development of prosthetic devices must be balanced against the realities of long-term disability, according to Hudspeth. Over the past 50 years, deaf people have advanced from a marginalized group to a community that can point to occupational and community successes. "Having achieved this success largely by their own efforts, some deaf people are understandably concerned that an electronic device such as the prosthesis might alienate their peers—or especially their children—from the deaf community," says Hudspeth.

Similar adaptations might reasonably be expected in instances when a prosthetic device helps restore sight to a blind person. More is involved here than simply making up for a deficit. By the time a visually impaired person reaches adulthood, he or she has accommodated to a world without sight. Activating a prosthetic visual system will require a fundamental reconfiguration of the blind person's brain. What's called for is a shift from a lifelong dependency on hearing and touch to the new "gift" of sight. Some hint of the difficulties involved in this transition can be gained from the experience of blind people who have recovered useful vision after successful eye operations (usually the removal of cataracts or scar tissue from the eyes). S. B., first described by English neuropsychologist Richard Gregory, is a particularly striking example.

After losing his vision at 10 months of age, S. B. was able to see again at age 52 following an operation at the Royal Birmingham Eye Hospital in England. A year later, he died in deep depression. While the operation had relieved him of his visual handicap, it also took away his peace of mind and self-respect. In his newfound identity as a seeing person, S. B. considered all of his accomplishments while blind (he had learned a trade, married, and fathered a family) as paltry, even pointless. Prosthetic devices must be designed and introduced so as to avoid similarly tragic outcomes.

———————————

Another approach to prosthetic devices involves the melding of neurons with silicon chips or other technological devices. But first scientists must overcome several daunting obstacles.

For instance, while neurons communicate with each other in milliseconds, computer processors send messages in nanoseconds—billionths of a second. But speed of transmission may be less important than the speed of internal processing. And here the neurons win hands down. A single neuron carries out some of its central processing in picoseconds—trillionths of a second. This near-instantaneous processing helps explain why

a computer will never equal the overall performance of a human brain.

With faster internal processing, the neuron can respond much more quickly to changes than a silicon chip, and therefore is less dependent on the speed of transmission from one neuron to the next.

Secondly, prosthetic devices will be limited by some of the simplistic ideas we hold about perception. Vision is much more complicated than neuroscientists believed only a few years ago. The brain of the macaque monkey possesses about 32 visual brain areas and more than 300 connecting pathways. Communication within the macaque visual system doesn't consist of relaying a visual message along a preset visual pathway. Instead, the connecting pathways go upwards and downwards, side to side. While specialized cells exist for the detection of certain visual aspects (color, movement, orientation, and such), each responds to some extent to other aspects. By averaging these additional responses, the brain builds up a richer and more complex visual repertoire than would be possible by simply combining the individual contributions of totally different and distinct cells.

———————————

In addition, the responses of individual neurons vary with attention, motivation, and past experience. Thus, seeing involves a variety of interlinked, crisscrossing influences that are poorly captured by descriptions emphasizing an orderly progression of signals proceeding like commuter traffic.

Integrating such designs into a computer network isn't proving easy. New discoveries about the brain's organization—such as the unanticipated complexity of the visual system—increase the challenge. As a result, neuroscientists are in the early stages of what may turn out to be an only partially achievable goal: welding together the separate domains of neurons and silicon chips.

One example of current efforts to connect a prosthetic device directly to neurons may serve as a harbinger. Reported in the professional journal *Nature Neuroscience*, it involved the implantation of electrodes in the brains of rats in order to monitor the activity of between 21 and 46 individual neurons. Previously the animals had learned to operate a robotic arm. Pressing a lever activated the arm to pick up a drop of water, while releasing the lever moved it towards the rat.

As the rats—now "wired"—repeated this exercise, the experimenters monitored their neuronal firing patterns. Eventually they discovered a pattern of firing from brain areas that are involved with voluntary movement. This occurred just moments before the animal reached out with its paw to obtain its water reward.

The experimenters constructed a device that could translate the newly generated neural signals into the language of computers. First they noted that many of the signals emanating from 32 neurons seemed most related to the pressing action. They then disconnected the robotic arm from the lever. At this point, the neural signals from those 32 neurons began driving the device. The rats had only to *think* about pressing the lever in order to activate it. In essence, thought or intention of movement could be separated from movement itself. Neurons now activated the robotic arm directly.

But since the rats weren't aware of the disconnection, they continued to press the now-useless lever. Eventually they caught on—and soon were slaking their thirst without moving a paw. The whole operation was now being carried out purely by thought, generated by activation of the proper neuronal circuit!

Obviously, similar thought-controlled devices—which could serve a prosthetic function—will be harder to develop for people. For one thing, human limb movements involve far more neurons and neuronal circuits. This increase in cell number and circuitry calls for greater precision in monitoring, along with more complicated prosthetic devices.

———————

Another difficulty may soon be overcome, however: the unsightly and cumbersome cables connecting the brain to any electronic circuitry. Biomedical engineers are already in the process of building brain implants that can communicate with a computer through radio signals, thereby eliminating the need for direct wiring.

For instance, a computer chip can be embedded in a muscle to detect the electrical impulses arriving from the brain and spinal cord to flex that muscle. That information can then be relayed to a computer that contains a program for opening a door or

■ FOLLOWING PAGES: Though only five years old, Jordan Adams is a piano prodigy who plays better than kids twice his age. When he practices a musical piece in his mind, he activates brain areas that are involved in imaging his finger movements. These areas—"turned on" by his imagination—surround the brain centers that spring into action when he actually plays.

moving a wheelchair whenever the muscle tightens. Devices are already in place that enable paralyzed people to move cursors on their computer monitors by thought alone. The electrical activity recorded by electroencephalogram (EEG) recordings is used to move the cursors. The patient has only to generate EEG activity to power the cursor. But so far, the technique has been only partially successful. The limiting factors are patient fatigue and imprecision in moving the cursor to the desired position.

Nevertheless, highly innovative prosthetic devices are likely in the next several years. To get a feeling for the challenges involved, look around you for a moment and then close your eyes. Reach outward towards something, relying only on your memory and your sense of your arm's position in space. Now compare your performance to how well you complete the same operation with your eyes open.

Obviously, visual search is easier and more accurate than relying on a mental map of your arm's position in space. When you reach for something, your brain takes the ease of visual search into account and programs your arm movements from the frame of reference of your eye. This makes a good deal of sense. In my own case, for instance, sitting at a rather paper-strewn desk, I have only to look at the desk and my brain will program smooth effortless movements around the different objects before me. But if my brain coordinates my arm's movement on the basis of that arm's position in space (instead of using visual guidance), a rather cumbersome translation is required. Each time I make one movement, my brain has to convert the visual scene into reference points involving my arm.

A computer-assisted prosthesis that was based on visual guidance would work differently. A device implanted in the brain would be used to scan the brain's impulses as it plans a movement, and then use these impulses to program the muscles to move the prosthesis.

But what about more elaborate and ambitious projects? Would a brain implant make possible the transfer of huge amounts of data from your brain to mine? Let's engage in a little thought experiment to test the implications. Imagine that you are an expert on, say, world affairs. Imagine further that after some futuristic placement of interconnected brain chips, all of the knowledge that you have accumulated over years on world affairs can now be transferred to my brain with absolutely no effort on my part. Do you think that would be a good idea? Actually such a brain-to-brain link would be an unwise and perhaps even dangerous intervention. For one thing, knowledge isn't embedded in the brain like groceries stacked on supermarket shelves. In fact, ideas and concepts aren't localized anywhere. Think back to Karl Lashley's attempt to discover the rat's "engram" for maze-running.

But there's another reason why such a direct transfer of knowledge wouldn't be such a wonderful development. Our senses serve the useful protective purpose of screening us from direct, unmediated contact with other people. Thanks to our eyes and ears and hands, all of us experience a "me" inside and an "everyone else" on the outside. A direct brain-to-brain prosthesis for ideas and knowledge would interfere with this delicate me/you balance that nature has established. Imagine how hard it would be to reliably distinguish your own thoughts and impressions from the contributions of other brains delivered to you via some interconnected brain chips. It's likely that all participants would experience alienation. Each would be deprived of that "warmth and intimacy" mentioned by William James as an integral part of self-experience.

In addition, this sort of sci-fi research isn't possible—because we haven't a clue about how the brain processes some concepts. For instance, where and how might a surgeon link a computer chip within the brain to communicate a concept like wisdom? Obviously it's much easier to augment the performance of a sensory organ or to program muscles to carry out simple routines.

Although nobody can predict the future applications of brain-computer research, the early experiments establish one extremely important principle: Brain cells have the capacity to directly activate prosthetic devices. Thus we can anticipate future technological developments aimed at boosting brain power and, in the process, bettering the lives of amputees and others with disabling injuries.

Finally, it's likely that future insights into the brain may come from unanticipated directions. This has certainly been the rule so far. Many of the earliest advances in psychopharmacology resulted from chance observations of the effects of chemicals on mood and behavior. A 1950s drug for tuberculosis, Iproniazid, was later used as an antidepressant, due to a chance observation of increased energy and improved spirits among tuberculosis patients who had been treated with it.

Numerous brain-altering drugs are derived from plants. LSD resulted from an attempt to alter the chemical properties of ergot, a highly poisonous mold thought to be responsible for bizarre epidemics of the Middle Ages such as St. Anthony's Fire, marked by delirium and hallucinations.

Plants also may provide novel insights into brain function and neural signaling, according to molecular biologist Frank Turano. When plants experience adverse conditions—insect attack, extreme temperature variation, heavy wind or rain—rapid and significant changes occur in their levels of GABA and glutamate, the same amino acids that serve as neurotransmitters in humans. Turano's research is aimed at deciphering how plants detect and respond to changes in their environment. This knowledge could help us understand similar systems in humans and other animals, and could impact the development of new drugs for illnesses such as Alzheimer's, epilepsy, schizophrenia, and depression, all of which have been linked to malfunctions of various neurotransmitter receptors.

Other benefits of plants: They're readily available, and far less controversial research subjects than animals. Also, several major pharmaceutical firms have formed consortiums with native peoples in remote forests and jungles, in a cooperative quest for new mind- and brain-altering plant species.

I believe our understanding of the mind and brain will continue to evolve in directions that no one can predict with confidence. That's because research on the mind and brain is a mix of science and art, calculation and serendipity. We have learned more about the brain in the past 50 years than we did in the previous two centuries, and it's likely that pace will continue to accelerate. Perhaps only a decade or two into the 21st century, today's knowledge of the mind and brain will appear embarrassingly elementary and tentative.

ANALYZE THIS

A nd now, let's do something different. In this final segment I'd like to convey some of the excitement that researchers and investigators experience as they plumb the mysteries of the mind. At first I thought I might interview specialists who study the brain at every level, from the molecular to the behavioral. But then I realized that such an approach really wouldn't accomplish my goal. The best way to give you a feeling for what it's like to study the human brain, I concluded, would be to invite you to participate in my own brain studies.

As a practicing neurologist and neuropsychiatrist, I study the brain in terms of what happens to a person when his or her brain malfunctions. People come to me—or are sometimes brought by concerned relatives or friends—because something in their behavior has changed for the worse.

Like Sherlock Holmes, I then set out to discover the cause of their changed behavior. And, like Arthur Conan Doyle's fictional detective, I make careful observations, ask artfully crafted questions, and—most of all—correlate what I hear and observe with what I have learned over the years. It's an exciting and always challenging task that never ceases to intrigue me. Most rewarding of all, of course, is the opportunity my work gives me to help other people.

In this chapter, I'd like to try to convey some of the excitement and challenge of neurology and neuropsychiatry by encouraging you to diagnose and treat several patients I've recently encountered in my office. I will present the basic facts about each of them; only their names have been changed. Your task will be that of the neurologist and neuropsychiatrist: to evaluate the patient and develop a plan of how you might proceed, based on what you've read earlier in this book.

Your goal is to reason your way toward a correct diagnosis. In order to guide your efforts, I'll ask questions at various points in the narrative. Whenever you come to a question, stop reading, and—before looking ahead at the answer—see if you can figure it out.

CASE ONE: On your schedule this particular morning are two new patients. The first, Ms. Stroop, is a 33-year-old business executive. Three days ago her right leg began "giving out" on her. The leg feels numb and she experiences difficulty maintaining her balance.

"It feels like what happens when you sit with your leg crossed for too long, only in this case the numbness hasn't gone away," she says. Questioning reveals that three months earlier she had temporarily lost vision in her right eye. She had also felt pain when attempting to move that eye. After five days, her vision cleared and the pain disappeared.

Today, as Ms. Stroop entered your office, you noticed that she walked in an uncertain and hesitant manner, as if she couldn't quite trust her right leg to support her. You examine the leg and find it weak. When you tap her right kneecap with a reflex hammer, the leg jerks suddenly outward—an abnormally increased reflex. You then tap the left knee and observe no hyperactivity there, while its strength is normal. You then look into each of your patient's eyes to observe the origin of the optic nerves, which connect to the visual centers in the brain. In the right eye, this area is pallid and washed out compared to the normal, pinkish-yellow tint that you observe in the left eye. As you conclude your examination, your patient suddenly breaks down and begins to cry. Until this moment she has been perfectly composed. She explains that she is frightened at the loss of control she has been experiencing. She wonders, "Is this whole thing just a response to the increased work stress I'm under at the moment? Do you think I would benefit from tranquilizers?"

At this point you must make a critical decision. Are Ms. Stroop's symptoms and your findings the result of a disease affecting some parts of her nervous system,

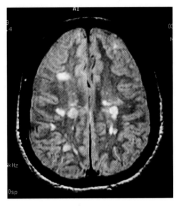

Fingerprint for MS: This CT scan of Ms. Stroop's brain (Case One) reveals light-colored globs known as demyelinating plaques—a sign of multiple sclerosis. Located deep within the white matter, they indicate that myelin is being lost from the nerve fibers. Depending on where they occur in the brain, they can cause weakness, numbness, loss of vision, or disturbances in balance.

or is she experiencing, as she suggests, a stress reaction? Certainly numbness in one leg and a sense of uncertainty while walking could result from stress. How do you decide?

Answer: You ask Ms. Stroop about her usual responses to stressful circumstances and uncover nothing in her background typical of a person with a low stress tolerance. The graduate of an Ivy League university, she is, in her own words, "a competitive person who thrives on the challenges offered by my job." What other reasons do you have for concluding Ms. Stroop's illness is not of psychological origin?

Answer: Your physical examination of Ms. Stroop revealed objective signs of neurological disease: The right leg is weak and its reflexes are abnormally exaggerated. In addition, her right eye shows signs of previous damage. This corresponds to the history she gave you of a temporary loss of vision in that eye three months earlier. You make your initial critical determination, concluding that Ms. Stoop isn't experiencing a stress reaction but a neurological illness.

Your next task is to diagnose that illness accurately and quickly. Has she experienced a stroke?

Answer: While the symptoms are similar to what might be expected for a stroke victim, Ms. Stroop's age provides an important clue. Strokes and other forms of vascular disease are uncommon in a woman of 33. Much more likely at her age is an illness resulting from an attack on the insulating sheath that surrounds neuronal axons. Are there hints of just such an attack in her description of previous symptoms?

Answer: The mysterious loss and recovery of vision in her right eye three months ago is consistent with inflammation and temporary dysfunction of the myelin sheath surrounding the optic nerve. When this happens, the nerve acts like an electrical cord deprived of its covering: It either fails to conduct electricity entirely, or its conductance is much reduced. In the optic nerve, this

condition would be expressed as blurred or even absent vision. After the inflammation ceases, however, the nerve grows a new myelin sheath, allowing normal vision to return. Indeed, Ms. Stroop's current symptoms regarding her right leg also might be explained by a demyelinating process—one that affects not her optic nerves but the nerves controlling that leg.

The most common demyelinating illness is multiple sclerosis (MS). An MRI of Ms. Stroop's brain shows plaques, or scars, in her left hemisphere (which explains the weakness in her right leg) and along her left optic nerve (which accounts for her temporary loss of vision three months ago). You tell her your diagnosis and suggest one of the recently available treatments that may arrest the course of MS. Most importantly, you provide her with reassurance, by informing her that many people with multiple sclerosis can lead perfectly normal lives.

CASE TWO: Your next patient is Jonathan Randall, a 49-year-old lawyer who comes to your office only because his wife insisted he seek help. For the first few minutes he doesn't say much and makes little eye contact. After some light social conversation, he relaxes a bit and tells you that over the past six months he has lost all enthusiasm for his law practice. "I don't even want to go into the office," he says.

Mr. Randall is also easily fatigued, loses his temper in response to petty annoyances, awakens frequently during the night, and can't concentrate or make decisions. On occasion he feels that "life is one hassle after another" and sometimes "isn't worth living"—but says he has never seriously contemplated suicide. A recent physical exam, he reports, was perfectly normal. Your questioning of Mr. Randall doesn't elicit any suggestions of neurological illness, and your physical examination of him doesn't turn up anything that points to brain dysfunction. Before telling him your diagnosis you ask Mr. Randall if he has any explanation for his feelings. His response:

"I'm depressed, I guess. I've tried to fight this thing on my own but I can't seem to do it."

Do you agree?

Answer: Mr. Randall's symptoms—his loss of enthusiasm for his chosen career and life in general, fatigue, poor sleep, inability to make decisions, and the change in his temperament—meet the criteria for major depressive reaction. His self-diagnosis is correct.

After answering some questions from him, you ask others of your own, including some about his medical family history. You discuss several antidepressants and convince him he might benefit by taking one. Question: Why was it important to ask Mr. Randall for his own thoughts as to what was happening to him? After all, he could have reasonably retorted: "If I knew what was wrong, I wouldn't be in here seeking your opinion."

Answer: Depression is a difficult diagnosis for many people to accept. Despite gains in society's acceptance of this disorder, a stigma remains associated with it and other forms of neuropsychiatric illness. Mr. Randall's self-diagnosis reveals that he is both insightful and not threatened by the diagnosis of depression. This suggests that he's likely to accept medications and make the necessary efforts to get well.

Three days later he calls to tell you he feels "jittery, wired, like my engine is going full throttle." His sleep is even more disturbed; during the daytime, he finds it difficult to sit still and feels the urge to pace around the house. "I'm worse now than when I came to your office," he complains." What's going on?"

What *is* going on?

Answer: Either the drug you prescribed is causing a switch from depression to mania in a man with an underlying bipolar disorder, or the changes Mr. Randall reports represent side effects of that medication. This is an important distinction to make, since the conversion of depression to mania can result in what's called a mixed state, in which the increased energy and agitation provoked by mania provide the depressed person with the will and determination to kill himself. A drug's side effect, in contrast, is much more easily managed. Which alternative is correct?

Answer: During your initial meeting, you considered the possibility of bipolar disorder. You requested Mr. Randall's detailed family history because you were looking for possible indicators for mania. Had anyone in his family ever been hospitalized or treated for an illness marked by mood swings, boundless energy, expressions of grandiosity, or other evidence of mania? Nothing in his background suggested mania or hypomania (a milder form of mania). Such a history makes bipolar disorder unlikely, and you conclude that Mr. Randall is experiencing overactivation due to the medication.

Antidepressants are primarily either activating or sedating. Since Mr. Randall initially complained of a lack of enthusiasm and an inability to motivate himself, an activating drug was the logical choice. But remember, he also mentioned that he lost his temper at petty annoyances. That suggests an element of overactivation. What do you do now?

Answer: You ask Mr. Randall to come to the office. After determining that his depression hasn't worsened and he isn't going to harm himself, you prescribe an antidepressant that can be depended upon to induce drowsiness. "Take this drug every night just before going to bed," you advise. "Call me in a few days and tell me how things are going."

A week later, Mr. Randall calls to say he is sleeping better, is less restless, and feels more motivated. He is on the road to recovery.

As discussed earlier, depression can take different forms in different people; the patient's symptoms largely determine your selection of antidepressant. Is the patient agitated or withdrawn? Is the depression worse in the morning or at night? Is it associated with sleeplessness or the tendency to sleep too much? Based on the answers to these and other questions, one antidepressant will be selected over another. In some patients—such as those afflicted with bipolar disorder—additional agents known as mood stabilizers will be added to prevent sudden mood swings.

Over the next three months, Mr. Randall regains enthusiasm for his work and reports that he no longer feels depressed. He then asks, "Can I stop taking the antidepressant now?" How do you respond?

Answer: In most instances, depression occurs in episodes and improves over time. As with any illness, medication can be stopped *when the patient is no longer sick.* But stopping an antidepressant is not something to be done cavalierly. As his neuropsychiatrist, you take several factors into account, such as whether this was a first episode, the seriousness of the depression (was the patient suicidal?), and whether the patient demonstrated any ability to recognize the depression and seek help. Mr. Randall scored well on everything but the last factor. If his wife had not insisted, he would have continued to "fight this thing on my own" and, as a result, his depression might have worsened to suicidal proportions.

Therefore, you tell Mr. Randall that he should continue taking his medication for now and that, during later discussions in your office, the two of you will determine a method for gradually tapering off the antidepressant. Discontinuance of the drug will be followed by monthly visits to make certain his improvement continues. Any evidence of recurrence can be promptly treated before the depression becomes disabling.

CASE THREE: When you return to your office after lunch, three new patients are on your schedule. The first is referred to you by a psychiatrist:

When Joan Holmes entered her apartment she "felt" that something was different. She quickly looked around and determined that nothing was missing. No signs of forced entry, no noticeable changes since she had left for the office that morning, other than…the coffee table. She spent several minutes staring at it.

"The table seemed strangely altered in a way I couldn't put my finger on," she tells you. "It was as if somebody had exchanged it for another just like it. But why would anyone do that? The coffee table wasn't anything special; I had picked it up at a sale at Danker's. Then a funny thing happened: After staring at it, my initial feeling disappeared and everything seemed OK. This was definitely my table."

A week later, Joan had another "strange experience," as she termed it. While talking to a friend of many years, "I suddenly had the feeling that this person was a stranger that I hardly knew at all. That feeling lasted a few seconds and then disappeared as quickly as it had come over me."

Two days after that, sitting in a movie theater, Joan "sensed" that she had seen the movie previously. But she knew the movie had just been released. At this point she was convinced that something odd was happening. Frightened and worried that she might be experiencing a nervous breakdown, she decided for the first time in her life to consult a psychiatrist.

After three sessions with Joan, the psychiatrist told her he thought the problem wasn't psychiatric but neurological in origin, and suggested that she contact you for an appointment. She did so, and—just two days before that appointment—she had felt "strange inside my body…like I was watching myself." Seconds later, she had fallen to the ground in an epileptic seizure.

When you meet Joan Holmes in your waiting room you encounter an anxious, 40-year-old computer specialist. After she tells you of her strange experiences you ask if she had ever believed, even for a moment, that someone had replaced or altered her coffee table.

"No," she says. "Nor did I ever believe I had really seen the movie before or that my friend was actually a stranger. It just seemed like these things were happening."

Why was it important to determine her attitude about the reality of those experiences?

Answer: Contrary to popular opinion, a delusion isn't diagnosed by weighing the likelihood of what a patient is telling you, but by assessing whether the patient is willing to consider alternative explanations. (Some delusions, particularly of persecution, could conceivably be true.)

Joan readily admits that she has no explanation for her experiences. Therefore, she isn't suffering from a delusion such as the belief that someone is controlling her by means of an influencing machine (a common schizophrenic delusion). Not only is she definitely not psychotic—out of touch with reality—but, as her psychiatrist correctly noted earlier, she shows no signs of any psychiatric illness.

At this point you ask her to elaborate on the words she had used to describe the times she had felt something was "different" or "strangely altered." You ask her to describe the sensation of having seen the movie before, the feeling that her friend was a stranger. You do this because these terms have something in common. What?

Answer: They all describe experiences. Each disturbance induces a thought ("Has somebody broken in to my apartment?") or an emotion (the anxiety and subsequent fear that she might be suffering a nervous breakdown). What part of the brain is concerned with experiences, and what do the symptoms tell us?

Answer: As mentioned in the Introduction, the temporal lobe plays a prominent role in integrating our inner experiences so as to provide us with our sense of self. Thanks to this lobe, we experience ourselves as essentially the same person from one day to the next— and the people and things we encounter on a regular basis seem familiar to us. Mild disturbances in the functioning of our temporal lobe include *déjà vu* (French for "previously seen," meaning a feeling that something brand new actually has been encountered previously) or

jamais vu ("never seen," a feeling of unfamiliarity when in familiar surroundings). Joan's experiences with the coffee table and her friend are typical *jamais vu* experiences; the movie was a *déjà vu* experience. *Déjà vu* is quite common; most of us have experienced it, usually under conditions of fatigue or stress. *Jamais vu* is less common. But if either or both of these experiences begin to increase in frequency and intensity, the neuropsychiatrist should begin to suspect a disturbance in the temporal lobe. What else points to a problem in Joan's temporal lobe?

Answer: Joan's epileptic attack, preceded by the *jamais vu* experience of feeling depersonalized, firmly points to the temporal lobe. Temporal lobe seizures are among the most common forms of epilepsy. You ask Joan if anyone has ever commented that she sometimes seems "dreamy," "not with it," or "in a trance." Does she ever miss parts of people's conversation? Does she forget things more often than previously? Why do you ask such questions?

Answer: At this point, you feel confident that Joan's problem emanates from a problem in the temporal lobe. Most likely she is experiencing brief seizures during which she loses contact with the people and events around her. Since the temporal lobe is intimately associated with the hippocampus—the site for the initial encoding of memory—you suspect that her memory may also be affected.

In response to your questions, she confirms that people do complain that she doesn't seem to be listening. And, yes, she *has* worried a lot recently about some "memory gaps." At this point, you order an MRI scan; it reveals an abnormal configuration of blood vessels in the left temporal lobe. A week later, Joan successfully undergoes neurosurgery to remove the abnormality. Her symptoms of "strangeness" permanently disappear.

CASE FOUR: Your second patient of the afternoon, a former construction worker named John Timmons,

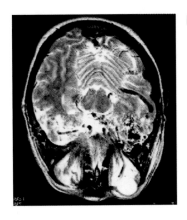

MRI of Joan Holmes's brain (Case Three) shows an abnormal network of blood vessels within her left temporal lobe. Since this lobe helps integrate personal experiences and provide a secure sense of self, disturbances here can result in symptoms that may be misdiagnosed as evidence of psychiatric illness. Fortunately, the abnormality was operable; after its removal, Joan's symptoms permanently disappeared.

arrives with his wife and a copy of his medical records. You begin reading them prior to going out to the waiting room to greet the couple, and discover that John was injured at work ten years ago, when a hydraulic jack struck him on the forehead and knocked him unconscious. Studies at the hospital revealed a fracture of the sinus abutting his left frontal lobe. Over the next several weeks John complained of difficulty in concentrating and in controlling his temper. His wife noticed that he seemed "out of it." She described him as changed and "harder to get along with."

"Between his flammable temper, irritability, and mood changes, I can't keep up," she wrote in a note to the neurologist that John had consulted six months after the accident. Given this information, what do you think was that neurologist's diagnosis?

Answer: John Timmons had suffered a concussion. Concussions often can be followed by temporary changes in concentration, mood, and temper control. Usually, the patient improves and returns to his or her previous personality.

But that didn't happen in this case. In fact, John got worse, exhibiting increased irritability and lowered frustration tolerance. The notes include a recollection by his wife of one instance when John started shouting at a stranger in a supermarket line because he had mistakenly concluded the stranger was trying to cut ahead. "I just couldn't believe it," the wife had said. "He didn't seem at all concerned about putting on a scene."

On another occasion, John displayed road rage—after a driver had "cut me off" in traffic, he recalled, he pursued the offending car and forced it to the side of the road, then started banging on the driver's window. The police arrived, and John narrowly missed being arrested.

"The other driver didn't want to press any charges," notes John's wife. "I think he could see the distress the children and I were experiencing. We'd been in that car and, believe me, he was driving like a madman. We all

thought we were going to be killed. I kept telling John that I didn't think the other driver had seen him and hadn't intended to cut him off. But he didn't pay any attention to me."

How would you characterize John's personality change? Injury to what part of his brain might explain this alteration?

Answer: Impulse control and the ability to put things in proper perspective are two important functions of the frontal lobes. John's previous neurologist had known this, of course, and had written, "Sudden flare-ups of temper and an inability to take an overview are John's greatest problems at the moment." He had ordered an MRI, which showed evidence of damage to the left frontal lobe, the same side that had endured the frontal sinus fracture. Most likely, John's head injury had resulted in a contusion, or bruise, of his left frontal lobe. Although the contusion had healed after several months, it had left a permanent scar.

You learn that John's life has been a long series of failures since his accident. He has not been able to keep a job, due to frequent arguments with co-workers that, on occasion, stop just short of physical violence. He has alienated neighbors and friends with his tendency to, in his words, "give them a piece of my mind." Previous unsuccessful treatments have included antidepressants, tranquilizers, mood stabilizers, even anti-epileptics. John briefly attended a head injury self-help group, but stopped going after several shouting matches erupted with other group members.

At this point you decide you have read enough and you bring John and his wife into your consulting room. You observe that John is a muscular man in his mid-30s who strikes you as suspicious and intimidating. He doesn't smile or extend his hand in greeting. "I don't know why I have to see yet another f_____ doctor," he says. "I've had enough of this s___." John is not here on his own. Instead, he is grudgingly following an insurance company directive that he be examined by an impartial neurologist to determine whether he can return to work.

You conduct an interview, and his answers are curt. When you point this out to him in the context of his tendency to become impatient and angry, he lets fly another string of expletives and raises his voice so loudly that your office manager buzzes you over the intercom.

"Everything's fine, Robin. We won't be but a few more minutes," you reassure her. Why is it reasonable for you to state that only a few more minutes will be needed to evaluate a problem that has existed for ten years, and what is your assessment?

Answer: Frontal lobe injuries are particularly tragic because they deprive a person, usually permanently, of their ability to appropriately evaluate and respond to people and events around them. In many ways, John's injury is similar to that of Phineas Gage, described in the Introduction. As with Gage and many others who have suffered frontal lobe injuries, John cannot separate the trivial from the important nor foresee the likely consequences of his behavior. His intensely hostile behavior toward you—a person he has never met before—indicates an inability to modulate his feelings despite possible adverse consequences.

He is well aware that you are in a position to recommend that the insurance company cut off his benefits and return him to work. Yet he is incapable—as he was during the episodes in the supermarket and on the highway—of modulating his anger and conducting himself in a socially appropriate manner. And, despite extensive (and varied) treatments, John has shown no improvement. He remains impulsive, angry, hostile, and uncooperative—and, sadly, is likely to stay that way. While you cannot predict that he will never improve, you can safely conclude that he is presently incapable of working at any job that brings him into contact with other people, and state so in your report to the insurance company.

CASE FIVE: Your final patient of the day is Barbara Tate, a 29-year-old stockbroker. While reading the intake sheet she filled out in the waiting room, you see that she wrote "headache" in response to the question, "What is the problem bringing you here today?"

Ms. Tate doesn't appear to be in any distress and answers your questions politely and succinctly. Her headaches, she says, begin with a feeling of being "out of sorts" and mildly disoriented. She tells you that, since she can't "think straight" at such times, she usually stops what she's doing and lies down on a couch in her office. She turns off the lights, tells her secretary not to bother her, and tries to take a short nap. If she succeeds in falling asleep she may be lucky and wake up feeling perfectly fine.

More often, however, the following sequence takes place: A pounding headache gradually builds up, accompanied by nausea and sometimes vomiting. After the headache stops she feels exhausted for several hours "as if I've been through a wringer."

The headaches first started when Ms. Tate was 16 years old; her family doctor diagnosed them as migraine. Over the past 13 years, they have occurred irregularly, remained generally mild, and responded reasonably well to over-the-counter prescriptions.

"It sounds to me like your family doctor was correct," you tell her. "Everything you've told me so far sounds like migraine."

At this point she tells you the real reason for her visit to your office. She is worried that she may have another brain disease in addition to migraine, something progressive and deadly like a brain tumor. She then tells you about an incident that happened a month ago in a supermarket. While writing a check for her groceries, she suddenly couldn't write anything but her name. Frightened and embarrassed, she signed the check and then awkwardly and hesitantly asked the clerk to fill in the rest. Driving home a little later, she felt disoriented and pulled over to the side of the road, where she sat through a "blinding headache" that lasted an hour.

"Looking back on it, the worst part was that I could barely get out the correct words to the checkout clerk while I was in the supermarket. Not only could I not write, I couldn't speak correctly either. From the clerk's facial expression I could tell that my speech must have sounded garbled. I can only assume he thought I was intoxicated," she says ruefully. What did Ms. Tate experience at the checkout counter and does this change your mind about the diagnosis of migraine?

Answer: She experienced aphasia, the inability to produce or comprehend language as a result of a disturbance in the language areas of the cerebral cortex (or in the white matter that connects the language areas). Since aphasia can result from brain tumors, strokes, and a host of serious brain diseases, it is not a symptom to be taken lightly. Certainly the fact that Ms. Tate suffers from migraine doesn't provide any guarantee she can't develop a brain tumor. What do you do now?

Answer: After puzzling out the challenges of the previous four patients, you've probably noticed that neurologists and neuropsychiatrists place a lot of emphasis on patient's histories. As one professor often told me, "Ask the correct questions and the patient will tell you what is wrong." And that is what you do. You ask Ms. Tate about her family and learn that her mother, two sisters, and a brother also suffer from migraine. You also find that no family member has ever been diagnosed with another brain disease, such as a tumor. How does this new information help you, and what does it imply about the aphasia?

Answer: Migraine runs in families. In Ms. Tate's case, her migraine is inherited from her mother. The fact that her siblings have migraine greatly increases the likelihood that everything Ms. Tate has told you is due to a migraine attack. Without a family history, migraine is a chancy diagnosis; conditions like the abnormal blood-vessel pattern found in Joan Holmes's temporal lobe often can mimic a migraine attack. In general, an aphasic episode is a troublesome sign. But *complicated migraine* can include disturbances in vision, feelings of weakness or numbness on one side of the body, or—as with Ms. Tate—transient speech and language disturbances. At this point, you're 99 percent certain that complicated migraine is the correct diagnosis. What might you do to raise that certainty even higher?

Answer: Ask Ms. Tate if she has ever had an MRI scan of her brain. If the answer is yes, the results were normal, and the test was carried out within the last few years, nothing more need be done. If she has never had such a test your next step is to order it and give your patient the assurance and peace of mind she deserves. As expected, the study is normal. You inform her of the results. Are you now absolutely certain that Ms. Tate's brain is harboring no other problems?

Answer: In medicine, you're never 100 percent certain that a patient doesn't have a particular illness—you can never prove a negative. For the anxious patient (and the equally anxious doctor!) there are always other tests that can be done. Eventually, all the safe ones get ordered; the remaining procedures often carry more risk than does the illness the doctor is attempting to rule out. The experienced physician doesn't put his patient at additional and unnecessary risk, expense, and inconvenience. So you tell Ms. Tate that, based on your knowledge and experience, everything she has told you is explainable by migraine. To cover the remote possibility that something else might be wrong, you tell her to call you in the event of any new developments.

Now that you have completed this book you should be in a position to understand and follow the exciting developments I expect will occur over the next decade, as neuroscientists and others continue to explore and solve the fascinating mysteries of the mind.

ACKNOWLEDGMENTS

The Book Division wishes to thank all of the individuals, institutions, and organizations mentioned in *Mysteries of the Mind* for their guidance and help. In addition, we are grateful to the following: Greta Arnold, Leah Bendavid-Val, William Bonner, Vicky Brandt, Patricia Campbell, Joe Carey, Judy Carte, Lyn Clement, Flora Battle Davis, Archibald J. Fobbs, Deanna Holzer, W. Breck Howard, Jennifer Hudin, Nathalie Humblot, Vlad Kharitonov, Rayford Kytle, Karen Loera, Mary McComb, Pat Murdo, Dawn Opstad, Paul Lionell Petty, Aleta Sarno, and Rebeca Townsend.

ABOUT THE AUTHOR

A leading authority on the brain, neurologist and best-selling author **Richard Restak**, M.D. has written many books on this subject, including *The Brain, Brainscapes,* and *The Infant Mind.* He has served on the advisory councils of the National Brain Tumor foundation, the Office of Interdisciplinary Studies—the Smithsonian Institution, and the Developments in Neuroscience Project. Restak also has been a consultant to PBS for its two series, *The Brain* and *The Mind,* and has appeared on NBC's *Today* show.

ADDITIONAL READING

Brain Facts: A Primer on the Brain and Nervous System (Washington, D.C.: The Society for Neuroscience, 1997)

Carter, Rita. *Mapping the Mind* (Berkeley: University of California Press, 1998)

Churchland, Patricia Smith. *Neurophilosophy: Toward a Unified Science of the Mind/Brain* (Cambridge, Mass.: The MIT Press, 1998)

Coren, Stanley. *Sleep Thieves: An Eye-Opening Exploration into the Science & Mysteries of Sleep* (New York: The Free Press, 1996)

Damasio, Antonio R. *The Feeling of What Happens* (New York: Harcourt Brace, 1999)

Damasio, Antonio R. *Descartes' Error: Emotion, Reason, and the Human Brain* (New York: Putnam, 1994; Harper Collins, 1995)

Ekman, Paul & Richard J. Davidson, eds. *The Nature of Emotion: Fundamental Questions* (Oxford University Press, 1994)

Gazzaniga, Michael S., Richard B. Ivry, George R. Mangun. *Cognitive Neuroscience: The Biology of the Mind* (New York: W.W. Norton & Co., 1998)

Gilling, Dick and Robin Brightwell. *The Human Brain* (New York: Facts on File Publications, 1982)

Glynn, Ian. *An Anatomy of Thought: The Origin and Machinery of Mind* (New York: Oxford University Press, 2000)

Kimura, Doreen. *Neuromotor Mechanisms in Human Communication* (New York: Oxford University Press, 1993)

Kotulak, Ronald. *Inside the Brain: Revolutionary Discoveries of How the Mind Works* (Kansas City: Andrews and McMeel, 1996)

Lavie, Peretz. *The Enchanted World of Sleep* (New Haven: Yale University Press, 1993)

Pert, Candace B. *Molecules of Emotion: The Science Behind Mind-Body Medicine* (New York: Touchstone, 1997)

Restak, Richard. *Brainscapes* (New York: Hyperion Books, 1995)

Restak, Richard. *The Brain* (New York: Bantam Books, 1984)

Time-Life Books. *Mind and Brain: Journey Through the Mind and Body* (Alexandria, Va.: Time-Life Books, 1993)

Wade, Nicholas, ed. *The Science Times Book of the Brain* (New York: The Lyons Press, 1998)

ILLUSTRATION CREDITS

INDEX

Illustrations are indicated by **boldface.**

MYSTERIES OF THE MIND

By Richard Restak

Published by the National Geographic Society

John M. Fahey, Jr., *President and Chief Executive Officer*
Gilbert M. Grosvenor, *Chairman of the Board*
Nina D. Hoffman, *Senior Vice President*

Prepared by the Book Division

William R. Gray, *Vice President and Director*
Charles Kogod, *Assistant Director*
Barbara A. Payne, *Editorial Director and Managing Editor*
David Griffin, *Design Director*

Staff for This Book

Tom Melham, *Editor*
Vickie Donovan, *Illustrations Editor*
Lyle Rosbotham, *Art Director*
Diana L. Vanek, *Researcher*
R. Gary Colbert, *Production Director*
Lewis R. Bassford, *Production Project Manager*
Sharon Kocsis Berry, *Illustrations Assistant*
Peggy Candore, *Assistant to the Director*
Deborah E. Patton, *Indexer*

Manufacturing and Quality Control

George V. White, *Director*
John T. Dunn, *Associate Director*
Clifton R. Brown, *Manager*
Phillip L. Schlosser, *Financial Analyst*

The world's largest nonprofit scientific and educational organization, the National Geographic Society was founded in 1888 "for the increase and diffusion of geographic knowledge." Since then it has supported scientific exploration and spread information to its more than nine million members worldwide.

The National Geographic Society educates and inspires millions every day through magazines, books, television programs, videos, maps and atlases, research grants, the National Geography Bee, teacher workshops, and innovative classroom materials.

The Society is supported through membership dues and income from the sale of its educational products. Members receive NATIONAL GEOGRAPHIC magazine—the Society's official journal—discounts on Society products, and other benefits.

For more information about the National Geographic Society and its educational programs and publications, please call 1-800-NGS-LINE (647-5463), or write to the following address:

National Geographic Society
1145 17th Street N.W.
Washington, D.C. 20036-4688 U.S.A.

Visit the Society's website at www.nationalgeographic.com.

Composition for this book by the National Geographic Society Book Division. Printed and bound by R. R. Donnelley & Sons, Willard, Ohio. Color separations by NEC, Nashville, Tennessee. Dust jacket printed by the Miken Co., Cheektowaga, New York.

Library of Congress Cataloging-in-Publication Data

Restak, Richard M., 1942-
    Mysteries of the mind / Richard Restak.
        p. cm.
    ISBN 0-7922-7941-7
    1. Neuropsychology. 2. Brain. I. Title.

QP360 .R456 2000
612.8'2--dc21
                                                00-027668